STANDARD
HANDBOOK
FOR
TELESCOPE
MAKING

STANDARD
HANDBOOK
FOR
TELESCOPE
MAKING

Revised Edition

Neale E. Howard

1817

HARPER & ROW, PUBLISHERS, New York
Cambridge, Philadelphia, San Francisco, London
Mexico City, São Paulo, Sydney

By the Author

THE TELESCOPE HANDBOOK AND STAR ATLAS

STANDARD HANDBOOK FOR TELESCOPE MAKING *(Revised Edition)*. Copyright © 1959 by Harper & Row, Publishers, Inc. Copyright © 1984 by Neale E. Howard. All rights reserved. Printed in the United States of America. No part of this book may be used or reproduced in any manner whatsoever without written permission except in the case of brief quotations embodied in critical articles and reviews. For information address Harper & Row, Publishers, Inc., 10 East 53rd Street, New York, N.Y. 10022. Published simultaneously in Canada by Fitzhenry & Whiteside Limited, Toronto.

Library of Congress Cataloging in Publication Data

Howard, Neale E.
  Standard handbook for telescope making.

  Bibliography: p.
  Includes index.
  1. Telescope, Reflecting—Design and construction—
Amateurs' manuals. I. Title.
QB88.H76 1983 522'.2 82-48120
ISBN 0-06-181394-X

84 85 86 87 10 9 8 7 6 5 4 3 2 1

Contents

Foreword

In the past few years there have been two important developments in amateur astronomy. One is the wide variety of mass-produced, but still expensive, telescopes now produced commercially, and the other the increasingly fine telescopes built by amateurs, ranging from simple Dobsonian types to complex equatorials employing many kinds of mountings.

The home-built instrument is still the choice of many amateur enthusiasts. This book is for those of you who would like to try building your own. What you create may not look quite as finished as the telescopes made by commercial manufacturers, but after all, the quality of the image seen through the eyepiece has nothing to do with the instrument's appearance.

The basic component of any Newtonian telescope is the mirror, a part the amateur can make as well as, and often better than, the mass producer. Although you will find directions in this book for making the other essentials, the eyepiece and diagonal mirror, these parts are best left to professional opticians because they require complex techniques and processes. Modern eyepieces are well worth their cost.

Today the tendency is toward ever larger and thinner mirrors, but these are very difficult for a beginner to make. For this reason the first-time mirror maker will do well to make a medium-sized, moderate focal length mirror, the 8-inch f/7 described in this book. The combination of such a mirror with a few good eyepieces will result in a fine all-purpose telescope. It will also give you the necessary experience if you should ever want to build a larger one. The chances are that you will be so well satisfied you may never want to change. However, in the event you succumb to the large-telescope fever, you may find it convenient to buy a big mirror (they're expensive) and build the rest of the instrument yourself. You will find a section in this book that tells you how.

I am indebted to many people for their help and advice in writing this book. Among them are Allan MacKintosh, founder and editor of the Maksutov Club, Gorton Carruth, former editor for Thomas Y. Crowell Company, and Robert E. Cox of *Astronomy* magazine. Then there are the many generous amateurs who have contributed photographs for this revised edition: Dennis Milon, Walter E. Hamler, John Sanford, Stephen Reed, David Ratledge, William Phillips, Michael Morrow, Thomas Britton, and Charles Weirzbichi. Most of all, my thanks to my patient and helpful wife, Barbara.

CHAPTER 1

On Telescopes in General

FEW PEOPLE can look through the eyepiece of a good telescope without immediately wanting to own one. Unfortunately a really fine telescope, even in these days when so many of them are on the market, is a luxury many of us cannot afford. The alternative—building one at home—is usually dismissed because of the reputed complexities and difficulties in the building of a precise optical instrument. Yet the truth is that anyone of ordinary skills and aptitudes can produce a telescope of breath-taking quality. The only qualifications you need are perseverance, patience, and a willingness to follow directions. If you have these characteristics (or can acquire them), you can build an instrument which will compare in performance with the best professional models and, incidentally, you will have a great deal of fun doing it. There are fifty thousand people in the United States who have already found this to be true.

As a hobby, telescope making is not expensive. Relatively few dollars will produce one which, when pointed at a planet or a distant star cluster, will report what it sees just as accurately and faithfully as a commercial product costing many times as much. Most amateurs build their first telescopes for economy and then, much to their surprise, find themselves at work on another, even though the first may have surpassed their fondest expectations. The fascination of constructing a precise instrument from simple equipment and inexpensive basic materials is a strong one. Those who succumb to this fascination become the most ardent telescope builders (known as T.N.'s or Telescope Nuts). For these people the completion of one telescope marks only a breathing spell before the beginning of a larger and more ambitious project.

What, actually, are the characteristics of a successful telescope builder? The first and most important is an even disposition and a willingness to spend time enough to do one part of the job right before rushing on to the next. Manual dexterity is helpful but not vital. A little information about tools and their use is also helpful but this can be picked up as you go along. In the same way the telescope builder will also acquire what technical knowledge he needs; little or none is necessary to begin with. The basic human requirements are those which most of us have anyway; it is not necessary to be a mechanical genius or a wizard at mathematics.

The cost will be perhaps $100 for a modest instrument and $150 for a more ambitious one, but the money goes out gradually and is therefore less keenly missed than if it had to be spent all at once. Even this amount can be reduced if the beginner is part of a group that can purchase materials in larger quantities with consequent savings. Telescope making is a gradual process where each part is completed before you go on to the next one. There is plenty of time to think about what must be acquired, whether the next item on the list must be purchased or can be made. Because of this, the budget can often be stretched to the point where the actual dollar outlay becomes pleasingly small.

Presumably each reader would like to build a telescope but is a little nervous about beginning. This book is written especially for you. We shall go forward in our project step by step, explaining everything as carefully as possible and attempting to anticipate your questions. Let's, then, turn to the first question of all.

WHAT KIND OF TELESCOPE SHALL WE BUILD?

There are three general types of telescopes: refractors, reflectors, and a combination of the two called catadioptric systems. Each is based on the same principle, which is to gather light from an object, bend the light to a point called the focus, at which an image of the original object is produced, and magnify this image until it can be seen by the human eye. That part of the telescope which gathers and focuses light to reproduce the original image is called the objective; the part which magnifies the image is the eyepiece.

In the refracting telescope, the objective is a lens placed at the front of the tube. Light passes through the lens and in so doing is bent inward until it comes to a focus at the far end of the tube. The eyepiece picks up the image at this point. In principle this is very simple, for all you need to do to produce such a telescope is to grind and polish a piece of glass on both sides until it becomes a lens, place it in one end of a tube, use another lens for an eyepiece, and start observing the heavens.

But you would be disappointed in the result, for the image would be blurred and would have fringes of color surrounding it. The difficulty is that a single lens not only bends ordinary light unevenly but also breaks it up into its component colors. You could correct this fault by using two lenses for your objective, each having a different curvature and made of a different kind of glass. But this complicates your task enormously, since you must grind four surfaces of glass instead of two. This is a formidable task even for one who has had considerable experience in shaping glass and would be discouragingly difficult for a beginner. Furthermore, since light passes *through* the lenses, they must be made of glass of exceptional clarity. Glass of this type is expensive. So we decide not to build a refractor just yet, and look around for a simpler type.

The reflecting telescope, instead of passing light rays through a piece of glass, bends them by reflection from the surface of a concave mirror. This means that you need grind only one surface of glass, which you now cover with silver or aluminum to make it reflective. Since no light passes through the glass you need not worry about its clarity or optical qualities. There is one complication, however. As can be seen from the diagram, the reflected rays of light get in the way of the incoming rays, unless you find some way of bending them out of the tube. Fortunately, there is a simple means of doing this, for you can interpose a small flat mirror, or diagonal, in the path of the reflected cone of light and attach the eyepiece to the side of the tube instead of the end. This cuts off some of the incoming light but not enough to be of serious consequence. Such an arrangement is the simplest and easiest kind of telescope to build, and it is the one we shall use.

Its lack of complications does not mean, however, that it is an inferior type, for quite the contrary is true. The largest telescopes in the world, the 200-inch Hale telescope at Mt. Palomar and the 120-inch at Lick Observatory, are both of this type. Furthermore, the professionals who made these California giants and the beginner in his basement workshop both use the same basic techniques and methods.

There are several other forms of the reflecting telescope but all involve the grinding of more than one optical surface. Chief of these is the Cassegrainian telescope, about which we will have more to say later.

Combination telescopes, which go under the imposing name of catadioptric systems, have an equally imposing optical arrangement. Like the reflectors, they have a mirror which acts as an objective, but they also have a correcting lens placed at the front end of the tube. Their great advantage is that they are compact and easy to use. The beginner is wise to steer clear of them for, as in the refractor, he must grind and polish several optical surfaces, each with its special curve. Telescopes of this kind range from the small commercial models now being produced to the photographic giants such as the 48-inch Schmidt used to supplement the work of the great California telescopes. The Schmidt telescope is not adapted to visual work,

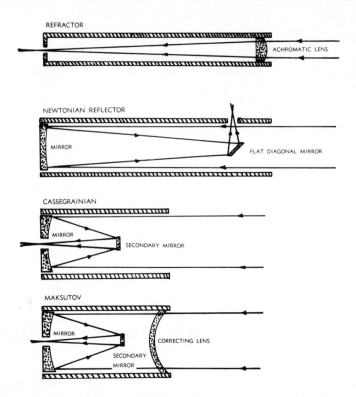

Schematic drawings of the path of light in telescopes. The eyepiece is placed at the intersection of the rays.

although a modification of the Schmidt design, the Maksutov, may be used both visually and photographically. We mention a few of these other telescopes because, after having completed some of the simpler types, you will be looking around for new worlds to conquer.

THE AMATEUR'S TELESCOPE

The simplest and most effective telescope we can make, then, is the reflector. As we have seen, it consists of three main elements: a concave mirror, a small diagonal flat mirror, and an eyepiece. Each must fulfill rather rigid specifications, while the remainder of the instrument serves only as a support so they may do their work well. There are as many varieties of mounting accessories as there are people who make them, yet it is this very lack of specific form that makes the construction of the other telescope parts as much fun, in a different way, as the more precise optical parts. Here is an opportunity for the imagination and ingenuity of the telescope maker to run wild, as we shall see when we come to the chapter on mountings.

Let us concern ourselves now with the essentials. If there is one part of the telescope which must be as nearly perfect as we can make it, it is the concave mirror, which is often called the primary. It is here that the rays of light which will form the image are reflected back to the point where they can be examined by the eyepiece. If each ray is to be treated alike to maintain its proper position in the image pattern, then each part of the mirror must be uniform with respect to all other parts. For this reason we shall give what may seem like a disproportionate amount of space to the subject of making the mirror.

But first we must decide what the general specifications of the mirror are to be, since they will determine the dimensions of the telescope itself. In the first place, the mirror must be large enough to collect as much light as possible, yet small enough so that it may be ground, polished, and mounted within the capabilities of the worker. A mirror 12 inches in diameter collects 9 times as much light as one 4 inches in diameter and therefore produces an image 9 times as bright, other things being equal. But it is many more than 9 times as hard to make. Since we don't want to set ourselves an impossible task, we shall choose an intermediate size of 8 inches, knowing that even at this figure we shall end up with a telescope of astonishing capabilities. It has been the practice for beginners to choose a 6-inch mirror for their initial efforts, possibly because it is a convenient size to work with. Yet the 8-inch mirror is only a little more difficult to grind and polish, and the extra trouble in making it is more than compensated for by the fact that it has almost twice as much area. All other things being equal, it is the area of the mirror of the reflecting telescope which defines its real value in terms of resolving power, or the ability to produce detail in images. This is the reason why telescopes have become progressively larger through the years, finally culminating in the 17-foot monster at Mt. Palomar.

The mirror must also have a focal length commensurate with its diameter. The focal length is simply the distance from the mirror's surface to the place where reflected light comes to a focus, or the rays converge at a point. A long focal length produces greater magnification, depending upon the eyepiece used, but it also produces a dimmer image. Also, the longer the focal length, the longer the tube must be. Now if the tube is too long, two disadvantages result. It must be built and supported very strongly or its vibrations will nullify the good qualities of the mirror, and every inch of length makes this harder to do. Secondly, a long tube is very awkward to use. In a Newtonian reflector such as we are building, the eyepiece is placed at the front of the tube; if the tube is too long the eyepiece will be very hard to reach when the telescope is in a vertical position. Short focal lengths also have disadvantages, the most important of which is that a deep curve must be ground in the mirror. For reasons which will appear later this can be a source of great difficulty.

So we compromise on a focal length of about 56 inches, which will put the eyepiece in a convenient position for most people. A 56-inch focal length for the 8-inch mirror is convenient not only as far as the length of the tube is concerned, but also from the optical characteristics it produces in the telescope. To understand why, we bring in a term familiar to camera owners, the f-number, or focal ratio. This is simply the focal length divided by the diameter, and is expressed mathematically as f/d. In this case, the focal ratio is $56/8$, or 7, and we refer to the telescope as an 8-inch $f/7$. Telescopes with focal ratios lower than 6 or greater than 9 begin to develop certain unfortunate characteristics, both in the use of the instrument and in the details of manufacture. But not only are those with focal ratios of 7 or 8 excellent general-purpose instruments, they are also the easiest to make. Therefore, our choice of 56 inches for the focal length of the mirror turns out to be a good one from all viewpoints.

Once the mirror diameter and focal ratio have been decided upon, we turn our attention to the other aspects of the telescope. We discuss each of these in some detail in later chapters, but in the meantime we want only a general idea of what lies ahead, so we treat them briefly here.

After the mirror has been ground to curve, its surface polished, and refined to the desired accuracy, it must be mounted in a cell which will support it adequately. The cell is then mounted at the end of a tube which we shall either buy or make ourselves according to the state of our finances. At the other end of the tube, at a distance commensurate with the focal length of the mirror, a diagonal mirror is mounted on a spider support stretched across the tube to support the diagonal securely. You may wonder why this diagonal doesn't shut off part of the image. Each part of the primary mirror contributes to the image, so the effect of the diagonal is only to cut down on the amount of light which passes it and thus affects only the brightness of the image and not its composition. The diagonal mirror is a small one, carefully ground to a perfectly flat surface, and can be either made or purchased since its cost is relatively small either way. Some telescope makers use a right-angle prism in place of the diagonal.

The final item of the assembly is an adapter tube for the eyepiece, and of course the eyepiece itself. All these optical parts are carefully aligned, and the most important part of the job has now been done. All that remains is to mount the telescope tube on a *rigid* and *unshakable* support. These adjectives are emphasized, for a good optical system demands a mounting strong enough to support it without vibration of any kind.

There are two types of mounting, each with many variations, the altazimuth and the equatorial. The first permits motion in altitude (vertical motion) and azimuth (horizontal motion); hence its name. The best example of this type of mounting is a camera on its tripod; it can move both up and down and laterally.

The stars do not move in straight lines as they cross the heavens; they follow curved paths. For this reason both degrees of motion must be used in an altazimuth mounting if the telescope is to follow a star across the sky. This is very awkward for the astronomer since moving a telescope in this way is a difficult operation to do smoothly and continuously.

The equatorial mounting, like the altazimuth, has two degrees of motion, but it differs in that its main axis is tipped parallel to the axis of the earth. This main, or polar, axis is always aligned in a north-south position. To follow a star moving from east to west, the telescope need be turned only around the polar axis. The other axis, called the declination axis, takes care of adjusting the telescope for the apparent height, or declination, of the star. Since the declination changes very little, for reasons explained in

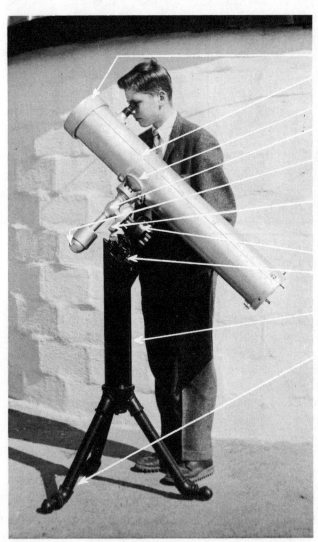

LEAD COUNTERWEIGHT: CAST IN MOLD MADE FROM TIN CAN

PIPE STRAP

TOP OF FIRE EXTINGUISHER

PIPE FITTINGS

COUNTERWEIGHT: MADE FROM IRON BALL

FRICTION CLUTCH: MADE FROM WHEEL DRUM OF LAWN MOWER

GEAR SYSTEM FOR SLOW MOTION: MADE FROM DISCARDED BLADE SHARPENER

VALVE HANDLE

CYLINDER FROM JUNKED FORCE PUMP

LEGS FROM ORNAMENTAL LAWN FURNITURE

As this photograph shows, the parts of a telescope can be made from a variety of sources.

a later chapter, the telescope needs to be moved in only one direction to follow a star across the heavens. This is a tremendous help in guiding the telescope for photographic purposes, although it must be admitted that it makes following a rapidly moving object such as a jet plane much more difficult. But since you are presumably building a telescope for astronomical purposes, the equatorial mounting is greatly to be preferred over the altazimuth.

This has been a necessarily brief description of the amateur's telescope. The details are given in later chapters. Nothing has been said here about telescope trimmings such as finders, setting circles, clock or synchronous motor drives, photographic accessories, or the like. These, too, are to be discussed later. The essentials of our projected telescope include the following:

1. It will be a simple Newtonian reflector.
2. It will have an 8-inch mirror whose focal length will be approximately 56 inches. In other words, it will be an 8-inch f/7.
3. In order to keep the cost as low as possible, we shall make all the parts ourselves, with the possible exception of the eyepiece and diagonal mirror.
4. It will have an equatorial mounting.
5. We shall have two fixed purposes during the entire operation: to create optical parts as nearly perfect as we can make them, and to provide a mounting whose sturdiness will permit our optics to do their job without interference.

SOME TELESCOPE HISTORY

Now that we have some idea of what kind of telescope we wish to build, it may be interesting to look back through the years to see how telescopes developed.

We think of the progress of modern astronomy as being due to the invention of the telescope, and to a large measure this is true. But much of the basic information about the heavens had been gathered and catalogued long before the first telescope made its appearance. Man has always studied the stars and speculated about their origin and nature, and during the sweep of the centuries he has piled up a mass of knowledge without benefit of optical aid for his observing eye. The planets and their apparent motions, the wanderers of space called comets, the groups and patterns of the stars, the moon, the earth itself in relation to its neighbors in space, all are as much a part of recorded history as man himself.

Some of this information was uncannily accurate. Eratosthenes, born 1700 years before Columbus, not only demonstrated that the earth is a globe but measured its circumference with a precision that is astounding.

This remarkable man also estimated the size of the sun and the moon and worked out their approximate distances from us. On the other side of the earth, the Maoris of New Zealand knew about the rings of Saturn and Jupiter's moons, and passed this information down through the centuries as part of their legends. Yet there is no mention of anything like a telescope in their history. How they knew this is one of the unsolved mysteries of antiquity, for biologists tell us it is physically impossible for the human eye to discern anything as faint as Saturn's rings.

Like many great inventions, the telescope came into being through a happy accident. So little is known of its inventor that historians are not even certain of the correct spelling of his name.

Jan Lippershey was a spectacle maker who lived in Middelburg, Holland. At some time during the year 1608, Jan's children, who had apparently been playing with some of his less valuable spectacle lenses, came rushing in to tell him they had found a way to bring a distant church steeple so close that they could see birds nesting under the spire. Like many fathers, Jan probably told them somewhat testily he didn't want to be bothered, and, again like many fathers, he followed them out to investigate. He found that they had lined up two of his lenses a foot or two apart in such a way that they were trained on the steeple. To his credit, he realized the importance of what he saw; and, from that day on, spectacles were ignored in the fascination of making various combinations of the new device. He finally became so expert in the art that he crowned his efforts at making simple telescopes with the production of a pair of workable binoculars. Lippershey's "magic tubes" became locally famous; and, even though he attempted to conceal his method of making them (in those days the patent laws were still in the far-distant future), word of the great new invention spread across Europe.

The incomparable Galileo Galilei, in far-off Italy, listened to descriptions of Lippershey's tube and proceeded to make one for himself. In common with the amateur telescope makers of today he wasn't satisfied with his first attempt. He made larger and longer tubes, using lenses of different characteristics and powers until he finally achieved a telescope which had a magnification of about 30.

Galileo's telescopes revealed features in the heavenly bodies which nobody had ever suspected were there. When he pointed them at the moon he was unable to believe that what he saw was really true, and was even more astounded at the appearance of the planets. Here was Venus with a chunk bitten from her side, Jupiter surrounded by four little disks which constantly changed their position, and Saturn looking sometimes like three planets and sometimes like a double-handled sugar bowl (this gives an idea of the quality of Galileo's telescopes; any amateur today can build better ones).

These heavenly wonders caught the imagination of the world; in order to observe them, men made telescopes longer and of ever greater magnification. Jan Lippershey's magic tubes grew until their inventor would not have recognized them. They were huge, ungainly contraptions which produced dancing, brightly colored images, but the discoveries made with them were tremendous and the seventeenth century became the golden age of astronomy. Dominique Cassini developed for the Observatory in Paris an "air telescope" which consisted of a lens hung from a tower and an eyepiece attached to the end of the long pole that guided the lens. With this tubeless device he discovered the fine black line that divides the rings of Saturn—and which is still called Cassini's division—as well as four of Saturn's moons and the dark belt line of Jupiter. The Dutchman Christian Huygens made a two-lensed eyepiece for his telescope and with it watched the polar caps of Mars and the equatorial bulge on Jupiter. Johannes Hevelius, a wealthy brewer turned astronomer, built a telescope 130 feet long for his observatory in Danzig. His maps of the moon are still famous, and he was the first to watch Mercury pass across the face of the sun.

Everywhere in Europe men were watching the skies, and in every astronomical workshop other men were searching for better ways to make telescopes. The giant devices used for observing fine detail on the moon and planets could be afforded only by observatories and wealthy men. The world was still looking for a telescope which might have universal use, for the small Galilean telescopes were far from perfect. (John Dollond, the English optician who finally worked out the principle of the achromatic [color-free] lens, in which two pieces of glass of different indices of refraction are used, was not to arrive on the scene until 1785, and this was still in the 1600's).

As is so often the case in scientific discovery, the next step was an idea looking for a man to point it out and put it to use. The man was not far away. Isaac Newton, only twenty-one years of age, had already decided that the refracting telescope had too many innate defects to make it useful and looked around for some other means of producing a really sharp image in a telescopic device. He found it in a paper published by the astronomer Nicolas Zucchi, in which the idea of the reflecting telescope was described. The result of Newton's work was the prototype of the telescope we have discussed earlier, and such instruments still bear his name, the Newtonian reflector. Since the art of coating a glass surface with a reflective material was not yet known, Newton ground his telescope mirrors and diagonals from a special metal alloy, still called speculum metal. The image formed was sharp and clear but not very bright since speculum metal is not as reflective as modern metal coatings.

In order to gather more light and produce a brighter image, men began to increase the size of the reflector, like that of the refractor before it. It culminated in such monstrous devices as the huge reflector of William Herschel, 4 feet in diameter and 40 feet long, and the even larger telescope of Lord Rosse of Parsonstown, Ireland. This one was so big that a man could walk down its 56-foot tube without bending his head, since the mirror was 6 feet in diameter. The quest for light-gathering instruments still goes on, although the ultimate in optical devices seems now to have been reached in the Palomar telescope, which dwarfs the greatest ones of the past.

While the Newtonian reflector was growing to its final stages, other types of telescopes were being evolved. Even before Newton perfected his reflector, the Scottish mathematician James Gregory had worked out an optical system which, although little used today, still deserves mention. Instead of using a plane mirror to transfer reflected rays outside the tube as Newton did, Gregory conceived the idea of placing a small concave secondary mirror in front of the primary mirror in such a way that the reflected rays were bounced back through a hole in the primary. The difficulty was that special curves had to be ground into each one, a paraboloid in the primary mirror and an ellipsoid in the smaller one. Neither Gregory nor his contemporaries were able to produce such surfaces because of limitations in the grinding surfaces, or laps, upon which mirrors were then ground to curve. Sixty years later a mirror-maker named Short succeeded in producing the desired curves, and the telescopes which resulted proved that Gregory's design was a good one.

A more successful idea was that of G. Cassegrain, a French sculptor, who in 1672 designed a reflector similar to Gregory's except that the secondary mirror was ground to a convex hyperboloidal figure and the main, or primary, was to be a paraboloid. These curves were as difficult to form as those of the Gregorian and for two centuries little more was heard of the design. When the techniques of mirror making became more advanced, however, the Cassegrainian telescope came into its own. Today it is a "popular" instrument, used by observatories and amateurs alike. It has several excellent features. The secondary mirror not only increases the magnification but also shortens the length of the tube so that the telescope is compact, easy to transport, and relatively simple to mount. There are several varieties of curves which can be used for the mirrors, each easier to produce than the ones designed by the inventor. Both the Gregorian and the Cassegrainian telescopes have the name "compound" attached to them since they require two curved surfaces which complement each other. We shall have more to say about the Cassegrainian later because it is well within the capabilities of the amateur telescope maker.

During the years when the reflector was rising to its zenith as "the" telescope—a period which reached its climax with the great William Herschel—the refractor was waiting for someone to do it the same service that Newton had done for the reflector. In the early 1800's such a man appeared. His name was Joseph von Fraunhofer, and his contribution to the art of telescope making was that he learned how to produce perfect optical glass. Up to this time it had been impossible to cast glass into disks of any size without introducing flaws and streaks throughout the material. These defects were deadly faults since they caused all sorts of weird effects in the refractor images. The only way of getting rid of these aberrations was to use lenses of long focal length, thus producing cumbersome and unwieldy instruments. Fraunhofer changed all this, for his disks produced images of sharp definition and wide field. The reflector now had an imposing rival.

The diameter of refractors now began to increase, and by the 1880's Alvan Clark, an American, was making the mightiest lenses the world has ever seen. These lenses are still in use all over the world. They include such classics as the 40-inch at Yerkes, the 36-inch at Lick, the 30-inch at Poulkova, and many others. Unfortunately, the size of objective lenses for refractors has a limiting factor since the only means of supporting them is to place mounting brackets around their rims. Their own weight, as they become larger, introduces distortion from sagging and bending. It is not surprising, because of this defect, that the 40-inch Yerkes lens built sixty years ago has only twice been surpassed in size. The giants in this field today are the Paris horizontal telescope and the 48-inch correcting lens for the Schmidt telescope at Mt. Palomar, California.

Reflectors are not subject to this limitation because they can be placed in cells which give them an evenly distributed support. The only factor which prevents reflectors from becoming ever more huge is the nature of glass itself. It now seems practically impossible to cast a glass disk of over 200 inches that will not crack or become otherwise distorted during the cooling period. But the advance of scientific knowledge may come up with an answer to this problem, as it has to others where the problem can be recognized and identified. If this is to happen, the world may one day see a 30-foot reflector. Such a day will be an exciting one in astronomical circles, for a telescope of this size is potentially capable of revealing the disk of a star other than the sun, something man has never seen. All telescopes now in use can do no more than record a star as a point of light, even though there are other means of determining its size and distance.

Refractors and reflectors went their own ways for three centuries before anyone thought of combining the two into a single telescope. Each type of telescope has its own advantages and difficulties. The refractor can give a wide field of view and a bright image and, if there are enough components in its objective lens, also yield a color-free image. But there must be a high degree of skill on the part of the lens maker and the glass must be flawless. This combination makes for an expensive telescope element. The reflector gives a beautifully flat and color-free image and is relatively easy to construct. But its field of view and its definition are sharply limited. This means that star images near the center of the field are sharp and clear, but those near the outer edge suffer from an umbrellalike distortion called coma. Coma increases with decrease in focal length. Below f/6 it becomes a problem. It is not a serious defect when the telescope is used for visual work, but for photographic purposes it is definitely a limiting factor.

In 1928 the German astronomer Bernard Schmidt successfully combined the two by placing a thin correcting plate in front of the main mirror of a reflector, and an entirely new type of telescope was born. Schmidt used his new device for photographic purposes, but in the years which followed his invention, modifications of design were made which permitted the system to be used visually. Such telescopes are called catadioptric systems and they combine the good qualities of both refractor and reflector. They are difficult to make but are not beyond the powers of the interested amateur. As a matter of fact, the first Schmidt telescope ever built in the United States was constructed by an amateur only two years after Bernard Schmidt's announcement of his great discovery.

A SUMMARY

We have attempted to demonstrate in this chapter that there is nothing mysterious about the way a telescope works. There are several points which should be re-emphasized.

1. Almost anyone who wishes to may build a fine telescope at small financial cost provided he has ordinary skills and is willing to put some thought and effort into the work.

2. A simple Newtonian reflector is ideally suited for the efforts of the beginner.

3. A quick survey of the history of telescopes shows that the great contributors to the field all started as amateurs. Even now the essential difference between the amateur and the professional is not in the quality of the product but that one group works for fun and the other as a vocation.

But we have talked about telescopes long enough. Now let's roll up our sleeves and make one.

CHAPTER 2

The Mirror: General Procedures

and Equipment

Since the mirror of a reflecting telescope is the key to its successful performance, we must consider the steps of its construction in detail. We shall understand each step much more completely, however, if we take a look first at the over-all procedures of mirror making and at the supplies and equipment we shall need. After this we can go back and discuss each step in point-by-point detail.

DEPTH OF CURVE

The first operation is to rough out a curve in the mirror. The curve must be spherical in shape, that is, it must correspond to an arc of a circle. But how deep must it be to produce the focal length we wish? Let's make use of a few facts from geometry. We recall that the amount of curvature of a circle is inversely proportional to the length of the radius: the shorter the radius, the deeper the curve will be. In optics this relationship is called the radius of curvature. Now if we place a light at the center of the circle of which the curve on the mirror is an arc, rays from the light will be reflected back to the center, as shown in *a*. In *b* we have moved the light so far from the mirror that its rays have become parallel, like those of a star. These rays, instead of being reflected back to the center of the circle, meet at a point, or focus, whose distance from the center is equal to one-half the radius. In other words, the focal length of a spherical mirror is equal to half its radius of curvature.

In diagram *c,* we show the mirror as part of a circle. The shaded area represents the amount of glass which must be removed from a flat surface to produce the curve shown. The depth of this curve at its central point is called the sagitta. The diagram shows that the sagitta would be deeper if the mirror were wider; therefore there is also a connection between the depth of curve (sagitta) and the radius of the mirror itself.

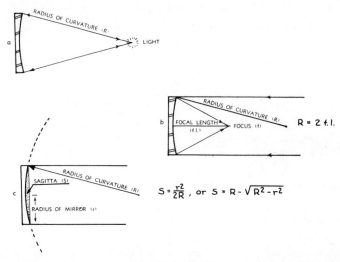

Relationship between radius of curvature (R), radius (r), focal length ($f.l.$), and sagitta (S), in reflecting telescopes.

Fortunately there is a simple formula which combines all the three variables to tell us how much glass to remove for our desired focal length.[1]

$$\text{depth of curve (or sagitta)} = \frac{r^2}{2R},$$

where r equals the radius of the mirror (in our case, 4 inches) and R equals the radius of curvature of its spherical surface.

For an 8-inch mirror whose focal length is to be 56 inches we must therefore hollow the glass out to a depth of .0714 inch (approximately $\frac{1}{14}$ inch) at its center.

$$\text{sagitta} = \frac{4^2}{2 \times (2 \times 56)} = .0714$$

But how to produce this hollow? The process is the fundamental operation of mirror making. If one piece of glass (the mirror) is ground on top of another (called the tool) with an abrasive between them, the uppermost

---

[1] The formula given above represents a very close approximation for the depth of curve of a spherical surface. The true formula is $S = R - \sqrt{R^2 - r^2}$. For mirrors whose focal ratio (see page 6) is f/5 and below, the true formula must be used. But for mirrors above f/5, $r^2/2R$ will give a value for sagitta so close to the true one that there is little point in using the more complex formula.

piece becomes hollow ground (concave) while the lower one acquires a convex surface. In practice, the grinding operation consists of pushing the mirror back and forth across the tool, using long strokes and changing direction periodically so that all parts of the mirror and tool surfaces are abraded. In this operation the center of the mirror is ground down faster than its edges. This is because the center maintains constant contact with the tool, while the edge on one side loses contact on every stroke because of its overhang. On the other hand, the edge of the tool is subject to more abrasive action than any other section since the center of the mirror exerts great pressure there.

We fasten the tool to the top of a barrel (so that we may walk around it and change the direction of the strokes), sprinkle Carborundum powder and water on it, and grind the mirror on this surface until we have reached a sufficient depth of curve. We can measure this depth (sagitta) by laying a ruler across the mirror and inserting objects of known size under the ruler until one of the right dimension ($\frac{1}{14}$ inch) just fits. Or we can make a template by inscribing on a sheet of cardboard or soft metal a curve of the same radius of curvature as we wish for the mirror and cut along the line. When this curve just fits the curve in the mirror, we have reached the desired depth. Accuracy in either of these methods is not of paramount importance since there are other means by which the depth can be checked more accurately. They are useful for quick estimates of sagitta so that we can get on with our work. If we have done the grinding carefully, the curve will be spherical in shape, and we are ready for the next step.

FINE GRINDING

At this point we have a rough-ground mirror but one that is useless for optical purposes until we have smoothed and refined its surface. We accomplish this by continued grinding, using finer and finer grades of abrasive until the rough surface has been transformed into a smooth one, free of pits and scratches. Since we no longer wish to *deepen* the curve, but merely to *smooth* it, we change the length of the stroke to a much shorter one. We also periodically change the position of mirror and tool, grinding approximately half the time with the tool on top instead of on the bottom. Reversing the mirror and tool in this way keeps the depth of curve constant. When the mirror is on top, the curve becomes deeper; when the tool is uppermost, the curve becomes more shallow. After we have completed work with the finest abrasive we shall have a mirror which still retains its spherical surface but which is *ground,* like the view finder in a camera, instead of *polished,* like the camera's lens. During the period of fine grinding we adopt another method of testing the curve, since we now have a surface so smooth that mechanical testing is useless. We discard the template and use a new testing device which makes use of the shadows found in a cone of light reflected from the mirror's surface. We discuss this method later.

POLISHING

At this stage we make what is called a pitch lap by pouring a layer of melted pitch over the surface of the tool and then cutting or pressing channels in it. We change from abrasives to polishing agents such as rouge or cerium oxide, and start the polishing procedure. This is done in much the same way as grinding except that we use a wider variety of strokes, are much more careful, and stop to test the surface much more often. We are now concerned not only with the general curve on the surface of the mirror but also with very minute deviations from this curve.

FIGURING

After we have removed all the minor imperfections from the surface of the mirror, we must transform the curve from a sphere to a paraboloid, since a spherical surface will not bring parallel light rays to a sharp focus. A paraboloid is the figure produced by the rotation of a parabola around its axis. On our mirror, it will be slightly deeper in the center and slightly shallower at the edges, than the sphere. But this is a fundamental difference for it transforms the mirror into an extremely precise optical instrument. The actual work performed in changing the curve from one form to another is very simple and need not concern us now except to note that it is done by changing the polishing strokes on the same pitch lap we have been using. In this stage we spend much time in measuring zones on the mirror's surface to be sure that we have produced a true paraboloid from edge to edge.

ALUMINIZING

After we are satisfied that the mirror is as nearly perfect as we can make it, we send it off to an optical house to have a coat of aluminum placed on its surface. This process requires too much expensive equipment for the amateur to attempt by himself. However, an alternative is to silver the surface of the mirror, a process which can be done at home. (See appendix for details.)

BASIC MATERIALS

(See appendix for a complete list of suppliers.)

Mirror Kits

All the materials we need to complete a precision mirror may be obtained in kit form from a number of dealers. An 8-inch kit costs from $35 to $50, the variation in price being due mostly to the number of abrasives as well as "extras" included in the kit. A good kit should include the following items:

Mirror Blank

The beginner may have his choice of Pyrex or plate glass for the mirror. The Pyrex costs a little more but it is preferred because of its low coefficient of expansion (resistance to changes in form because of temperature differences), its toughness, and its resistance to breakage. An 8-inch blank should be at least 1⅜ inches thick. The thickness of the blank is very important since we wish to take no chance that the fine curve we place on its surface will not be deformed by flexure of a too-thin mirror. These Pyrex blanks are usually made in the correct proportions of width to thickness, but if there is any doubt, the table below will be useful.

Tool Blank

This may be Pyrex, but need not be. Some amateurs have found plate glass tools preferable to Pyrex since the plate glass yields a smooth, evenly ground surface and is easy to work with. For an 8-inch tool, the glass should be 1 inch thick. Some dealers supply ceramic tools instead of Pyrex or plate glass. Although these are a little cheaper, they are very satisfactory and there are many mirror makers who prefer them.

MIRROR AND TOOL DIMENSIONS

Mirror Diameter (inches)	Thickness (inches)	Approximate Weight (pounds)
4¾	¾	2
6	1	3
8	1⅜	8
10	1¾	14
12½	2⅛	27

Tool Diameter (inches)	Thickness (inches)	Approximate Weight (pounds)
4¾	¾	2
6	⅝	3
8	1	6
10	1¼	9
12½	1¼	13

Abrasives

The kind of abrasive and the variety of grit sizes offered will vary among dealers. The sizes listed below refer to the number of openings per linear inch in the mesh through which the abrasive has been graded. For example, No. 400 means that this size has been separated by passing it through a mesh of 400 openings per linear inch. It also means that no particle is *larger* than this size. Abrasives come packed in cardboard

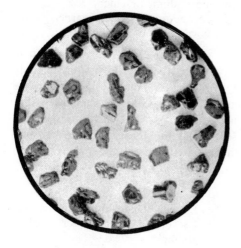

SILICON CARBIDE
(CARBORUNDUM)

ALUMINUM OXIDE
(ALUNDUM)

Two types of abrasives. Notice the difference in edge sharpness. This is why Carborundum is used for roughing out the curve and aluminum oxide for fine grinding.

boxes, tins, shakers, or plastic bags. If any of the containers has been broken open in transit, the whole shipment should be returned to the dealer, since the small grit sizes may have become contaminated with larger ones. A kit should contain the following grit sizes and quantities:

16 oz. No. 80 Carborundum
8 oz. No. 120 Carborundum, Alundum, or Aloxite
2 oz. No. 220 Carborundum, Alundum, Aloxite, or corundum
2 oz. No. 280 Alundum, Aloxite, or corundum
2 oz. No. 400 Alundum, Aloxite, corundum, or emery
2 oz. No. 600 Alundum, Aloxite, corundum, emery, or garnet
2 oz. No. 900 Alundum, Aloxite, corundum, emery, or garnet
2 oz. M305 emery

If you are bewildered by the variety of abrasives in this list, they simply represent various materials offered by dealers. It makes little difference whether you use No. 400 Alundum or corundum, for example. It is the grit *size* and how well the dealer has graded this size which is important, and the composition of the grit is secondary. For those who like to know exactly what they are working with, however, the table below will help.

COMMON NAME	CHEMICAL NAME	CHARACTERISTICS AND USE
Carborundum (Shortened to Carbo)	Silicon carbide	Extremely hard; cuts deep and fast. Use in the beginning for rough grinding.
Alundum Aloxite	Fused aluminum oxide; made in the electric furnace	Slower cutting action but does not leave pits as deep as Carborundum. Ordinarily used in intermediate stages of grinding, but may be used for all stages.
Corundum	Naturally occurring aluminum oxide	Like Alundum and Aloxite. Used for intermediate and final stages.
Emery	Mixture of Corundum, magnetite, hematite (iron oxides), and quartz (silicon dioxide)	Soft, smooth action. Used for fine grinding.
Garnet	Mixture of oxides and silicates of iron, calcium, and aluminum	Has sharper action than emery. Used in final stages of fine grinding.

Polishing Agents

Rouge (ferric oxide) is widely used and is supplied in all kits. It has the disadvantage that it polishes slowly and stains everything it comes in contact with: utensils, clothing, and hands. There are many amateurs who prefer not to use it. But for the final stages of polishing it has no equal. For preliminary polishing, however, most suppliers include cerium oxide, which not only polishes very rapidly but is clean to handle and easy to use. Another commercial product called Barnesite (a mixture of the rare earth oxides) is superior to both rouge and cerium oxide in rapidity of polishing, and some mirror makers think that it produces a surface equal to that by rouge. Since Barnesite is relatively expensive (about twice the cost of rouge), most dealers include only 4 ounces of rouge and 2 ounces of cerium oxide as standard items in their kits.

Pitch

Pitch has many uses in optical work, but it is important principally in making laps for polishing. There are several kinds. *Burgundy* pitch is prepared from the turpentine of the Norway spruce, while *Canadian* pitch comes from the hemlock. Each has a melting point of about 140° F. Coal-tar pitches are a product of the destructive distillation of coal and have a higher melting point, around 160°–170° F. The *kind* of pitch used is not very important, but the hardness is a vital factor in good polishing. Some of the pitch obtained from dealers must be diluted slightly with turpentine before it can be used. Its usefulness in polishing depends upon the fact that, although it appears to be a solid, it is actually a supercooled liquid and will flow to relieve any pressure placed on it. For this reason it adapts itself to the surface of the mirror, permitting very close contact to be obtained, and thus acts as an excellent base for polishing materials. Most dealers supply a half-pound bar with the standard kit.

Other Accessories

The list that follows, with one or two exceptions, is not included in kits. Most of the items can be made or bought as is very cheaply, and the whole list will not add materially to the cost of the mirror.

1. An 8×11-inch sheet of heavy cardboard or a sheet of $\frac{1}{32}$-inch aluminum of the same size. This will be used to make a template to measure the depth of the curve in the mirror.
2. A dozen salt shakers—from the dime store—in which to store the coarser abrasives. These should be labeled with adhesive stickers to show the size of the abrasive they contain.
3. Three or four plastic mustard bottles obtainable at the dime store. These are "squeeze" bottles, usually made of polyethylene, which have dispensing spouts. They are excellent for storing water suspensions of fine abrasives (No. 400 up) and polishing agents.
4. Three or four cheap cellulose sponges. Each should be cut into three parts. One of these parts is used to clean up after each size of abrasive and then thrown away—so that it cannot possibly contaminate other abrasives.

5. A supply of paper towels, and a half dozen dinner-size paper plates. A box or two of Kleenex.

6. A Carborundum stone, medium grit.

7. A Bunsen burner or small electric plate for heating pitch. The kitchen range will do nicely for this if the kitchen commandant permits.

8. A few ounces of turpentine for diluting pitch.

9. A lap-maker's mold. This is a rubber mold divided into 1¼-inch squares in a screenlike arrangement and is used to produce square facets on the pitch lap. They can be obtained from, among other dealers, the Edmund Scientific Company (see appendix), but it will be wise to read the chapter on polishing before you order one. You may decide to use something else.

10. A dime-store one-quart saucepan in which to melt pitch. Don't expect to use this pan for anything else. The Biblical injunction about pitch is very applicable here.[2]

11. A pan large enough to hold mirror and tool and enough water to immerse them in, with spacers between.

12. A sharp chisel or X-acto knife for trimming the pitch lap.

13. A few ounces of beeswax or the paste type of floor-polishing wax —the kind that contains carnauba wax as a base. This will be applied to the pitch lap.

14. A small paint brush (dime-store variety).

15. A yard of plastic window screening. This will be applied to the pitch lap and must be of the woven variety (*not* the kind where the threads are fused together) since it must stretch to conform to the lap.

16. A yard of plastic window-curtain or thin shower-curtain material. This is the kind that has minute projections embossed in its surface. This material is not really necessary, but as we shall see in chapter 6, it provides a welcome short cut in the polishing process.

17. An atomizer bottle. This kind used to spray window-cleaning fluid is very useful here.

18. A magnifying lens or, better, a positive eyepiece. (Use one from your binoculars.)

19. A support to hold the mirror upright during testing operations. Make it yourself out of two pieces of ½-inch plywood set at right angles to each other.

20. A Foucault testing device (to be described in detail later).

THE WORKSHOP

There are a few workshop conditions which will be most helpful in

[2] Ecclesiasticus XIII, 1.

speeding up the grinding and polishing processes and in making your work easier. The most important consideration is that a fairly constant temperature can be maintained. This is especially important during the testing and figuring of the mirror because even a slight change in temperature can seriously affect the accuracy of the work. Almost as important is that the workshop be as dust-free as possible. Dust is an excellent abrasive, as those who have gotten a few specks of it on a highly polished mirror have found. The scratches it produces on an otherwise flawless surface can be horrifyingly large and deep. If the dust-free requirement can be met, the corner of a cellar makes an admirable workshop, for temperature changes are not likely to be great. If your cellar doesn't

Raw materials. The tool (*center*) used for this 8-inch mirror is a ceramic material similar to pipe tile.

meet the dust-free requirement, you can improve the work space in this respect by building a plastic tent made from painters' transparent drop-cloths. These can be obtained very cheaply from mail-order houses or dealers in surplus stocks. Make the tent rectangular and large enough to work in—at least 8 feet square—by hanging the cloths from the cellar beams. Don't forget to cover the ceiling; dust dislodged from there by people walking around upstairs is even more pernicious than the kind that floats in the air. Spray the workshop with water from an insecticide sprayer a half hour before you start to work. The water vapor settles and takes dust with it.

The workshop must contain, as its principal piece of equipment, a grinding stand placed so that there is access to it from all four sides. A 55-gallon oil drum, stood on end and filled two-thirds full of water, makes an admirable grinding stand. If necessary, place small wooden shims under the bottom rim of the barrel so that there is absolutely no motion when lateral pressure is applied to the top. Using a punch and a heavy hammer, perforate the top of the barrel, making at least 30 holes in its surface. The purpose of these holes is to drain away water, grit, and any other products of the grinding and polishing operations. If the products of one step of the grinding process are flushed into the top of the barrel they are safely out of the way of subsequent steps. The coarse grains from the roughing-out steps of grinding will produce ruinous scratches if even one of them gets into the fine grinding or polishing stages. If left around on work surfaces they are sure to do just this. But once safely flushed into the barrel they are forever out of the way. If the barrel fills up after many flushings it is easy to siphon off some of its contents with a piece of rubber tubing and a bucket.

Cut a 12-inch circle out of a piece of $2 \times 12 \times 12$-inch material—it need not be hardwood; spruce or pine will do—and bolt it to the top of the barrel. Countersink the bolt heads and fill in the holes with plastic wood or wood glue. If you do not have access to a bandsaw to produce a circular work surface, use a square of 2×12 instead. Place 3 cleats at $120°$ intervals around the glass tool to hold it in position on the surface and you are ready to go to work.

We have used barrels for many years in our workshop and have found nothing better since they produce an unshakable support for grinding operations, are just about the right height for most workers, and make an ideal reservoir for waste materials. But their weight is objectionable unless they can be placed on a concrete or otherwise solidly supported floor. A substitute which weighs less, but still provides solidity, consists of a piece of 4-inch-diameter pipe set upright in a small tub of concrete. To the top of the pipe screw a standard flange, then bolt on a sandwich made of two pieces of $\frac{1}{2}$-inch plywood screwed together. The height of any grinding platform is very important to the comfort of the worker. When possible, it should be high enough so that the operator may work with a straight back, but low enough so that he may work the mirror back and forth across the tool without lifting his elbows. For most people this height is about 40 inches.

If either of the above is impossible to obtain, a very heavy table (or one held down with weights) may be used. The tool is cleated to the corner of the table. But, as we have said, the mirror must pass back and forth on all diameters of the tool; and, since we cannot walk around the table, we must make some arrangement so that the tool can periodically be rotated in its cleats. This sounds simple but in practice is difficult because we are likely to forget to rotate the tool or have trouble keeping the cleats in adjustment.

The workshop must be large enough to provide space for testing. In the Foucault test the mirror must be separated from the testing device by a distance equal to its radius of curvature, in this case, 112 inches. Consequently we must have a clear area at least eleven feet long. Also we must provide rugged stands to support mirror and tester. One of these can be the grinding stand and the other a very sturdy small table.

Finally, we need good illumination in the workshop and, conversely, a means of darkening the room to carry out tests. A storage cabinet or two in which to keep abrasives, polishers, tools, the mirror itself, and other equipment is worth its weight in gold. If these cabinets have locks, so much the better, since many a mirror has been ruined by casual visitors. We once had to re-do a 12-inch mirror because an unattended visitor wanted to see what the polishing operation felt like. His technique was fine, but he used No. 120 Carborundum instead of rouge. Perhaps the most desirable piece of equipment to have in the workshop would be a sink, but unfortunately this is a commodity usually not available in the ordinary cellar, and we shall probably have to carry our water in buckets.

The foregoing represents what we should *like* to have for an optical workshop and its contents. But it is rarely attained and, if impossible, we should not worry too much about it. Many excellent mirrors have been made in old sheds, a corner of the garage, attics, or any out-of-the-way place where a mirror maker can work in peace.

CHAPTER 3

The Mirror: Rough Grinding

Having made all our preparations and set up our equipment, we are ready to begin the first stage in mirror making: rough grinding.

As soon as the mirror arrives from the dealer, inspect its surface carefully. If it is Pyrex, grind its widest surface (these mirrors taper from back to front); if it is plate glass, grind its smoothest and flattest surface. Look for small bubbles just under the surface of the glass. It is almost impossible to cast Pyrex glass without bubbles, so do not worry if there are many of them scattered throughout the interior of the disk. But bubbles near the surface will turn into holes as soon as you grind down to them; for this reason a mirror having a bubble over $\frac{1}{16}$ inch in diameter which will become exposed on grinding should be returned to the dealer. These bubbles are easy to identify. You measure their width by laying a ruler on the surface and looking down through the glass. By looking through the *side* of the mirror you can estimate their depth below the surface. Remember that any bubble closer than $\frac{1}{16}$ inch to the surface will be exposed when you grind down to it. But then the bubble will be a hole.

Before starting to grind, bevel the edges of both mirror and tool, using a Carborundum stone (medium grade is best, for there will be less chipping than with coarser grades). It goes without saying that the edges to be beveled are only those on the surfaces to be ground. Unless you are careful, you will chip the edge of the mirror during the beveling process. Hold the stone at a 45° angle and direct your strokes *away* from the mirror. Keep both mirror and stone wet at all times. The bevel should

26

be about ⅛ inch wide. In the absence of a Carborundum stone, you can do a good job with wet No. 220 Carbo and a flat piece of metal. As any mirror is ground, its edge has a tendency to become sharp. Eventually a small chip will break off the sharp edge and, if it gets in between mirror and tool, will produce a deep scratch on the surface of the mirror. This is not serious during rough grinding but, later on when we have achieved a fine surface, it is catastrophic for it means we must go back to coarse abrasives to grind it out. Beveling the edges beforehand prevents all this.

After beveling is complete, attach the tool to the grinding platform on top of the barrel or grinding stand. Do this by placing the tool on the platform and tracing a circle around it. Then place three cleats at 120° intervals around the outside of the circle. The cleats should be below the level of the surface of the tool at the point where they touch the tool and should then slope away to a height of not more than ⅛ inch. Use 1-inch No. 6 screws to attach the cleats, angle them so that they tend to pull the cleats toward the tool, and countersink them to prevent the mirror's surface from coming in contact with a screw head. The cleats must not grip the tool tightly but produce just snug enough a fit so that the tool may be removed and replaced readily. Some authorities

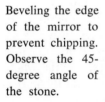

Beveling the edge of the mirror to prevent chipping. Observe the 45-degree angle of the stone.

permit a play of $\frac{1}{32}$ inch between tool and cleats, but most workers will find this too loose.

Place three sheets of wet paper toweling in the flat area between the cleats (this helps hold the tool in place and also provides a firm seat for it), place the tool on top of the wet paper, push it down firmly, and you are ready to begin rough grinding. During the early stages of rough grinding you don't really need the cleats because the wet paper keeps the tool from sliding. Later on, however, when the drag of the mirror across the tool becomes greater because of the finer abrasive, the cleats become necessary. They are essential during the polishing process, so you might as well have them from the beginning.

In the meantime place the abrasives in clean, labeled salt shakers or some other container which permits shaking out a small quantity of abrasive at a time. When you do this be careful to handle the fine abrasives first. This points up an essential precaution in handling abrasives; always be on the lookout for possible contamination of fine abrasives with particles of a coarser one. Be equally careful in washing-up operations. Use a small piece of cellulose sponge with each abrasive and store that sponge with its own abrasive. A good way to prevent contamination is to place the sponge in the bottom of a jelly glass or similar container, then invert glass and sponge over the salt shaker containing the abrasive corresponding to that particular sponge. In this way each abrasive is protected from all others and from dust as well. Finally, if abrasives are stored in a cabinet (preferably each one on its own shelf), always place polishing agents on the top shelf, then the abrasives in order of their increasing particle size until the coarsest end up on a bottom shelf.

Spray the surface of the tool with water from the atomizer, sprinkle a small amount of No. 80 Carbo on the wet tool and spray again. Just how much water to add is a matter of experience. Too much will make the abrasive wash away very quickly; too little makes it cake into a mud which cuts very slowly. The amount of abrasive added is important, too. There should be only enough to cover the surface of the tool *thinly,* never more than sufficient to permit each grain to occupy its own space. Abrasives work best when each particle can roll between tool and mirror. If too much is applied, the particles erode each other instead of the glass, and the whole quantity is broken up into a mud. We are anxious to hollow out the mirror as quickly as possible. This is done fastest when we err on the side of too little abrasive instead of too much.

THE FUNDAMENTAL STROKES

We are now ready to consider the fundamental strokes, or motions, of the mirror maker. Place the mirror on the tool and grasp it so that the fingers barely curl over the edge. This is to avoid transferring the heat of the finger tips to areas close to the surface being eroded and is a precaution to observe throughout all grinding and polishing operations. Believe it or not, the heat of your fingers causes the mirror to expand a little, thus producing a slight bump in its surface. The bump is ground off and, when the mirror cools, produces a hole. This is not important during the grinding stage, but during the polishing and figuring operations it is vital. So acquire the habit early of keeping the fingers away from the surface of the mirror, and it will save trouble later.

The "stroke" of the mirror maker is simply the act of pushing the mirror back and forth across the tool. At regular intervals he steps crabwise around the grinding stand, at the same time turning the mirror in such a way (described later) that a new diameter of the mirror travels across a new diameter of the tool. This combination of actions—pushing the mirror back and forth, walking around the barrel, and turning the mirror itself—will grind a hollow in the mirror and create a hump on the tool. The *way* it is done determines whether the hollow in the mirror is a smooth curve or just a hole. The amount of grinding which takes place in any part of the mirror depends on the size of the abrasive particles, the pressure applied, and the length of the stroke.

The cardinal principle of all grinding (and later on, polishing) is that the nearer the center of the mirror approaches the edge of the tool, the more rapidly the central part of the mirror is hollowed out. Conversely, if the center of the mirror moves only slightly away from that of the tool, the greater is the abrasive action on the mirror's edge.

The length of stroke is always spoken of in relation to the mirror's diameter. Thus a ½ stroke means that the center of the mirror moves a total distance equal to *half* its diameter: In an 8-inch mirror this would be 4 inches, adjusted so that the mirror overhangs the tool 2 inches at each end of the stroke, whether forward or backward.

TYPES OF STROKES

There are only four types of strokes, each having its own specific purpose. Although there are many variations of each type (each mirror maker works out his own favorite), these four are all that are needed to grind and polish any moderate-sized mirror.

The Chordal Stroke

The mirror is pushed back and forth along *chords* near the edge of the tool, as shown in the diagram. As you can see, the center of the mirror travels around the edge of the tool. The purpose of this stroke is to hollow out the center of the mirror rapidly, and the shorter the chord, the more drastic the action and the smaller and deeper the hole in the mirror's center. By grinding along chords nearer the center of the tool, the hole

Using the chordal stroke to hollow out the mirror. The mirror shown is a short-focus 6-inch, and the hollow is deep, as you can see from the inclination of the mirror.

This shows the lateral overhang at the beginning of the "W" stroke.

Diametral, or center-over-center, stroke.

in the mirror can be made to spread toward the edge. The chordal stroke is used only at the beginning of grinding, where your purpose is to remove quantities of glass as rapidly as possible.

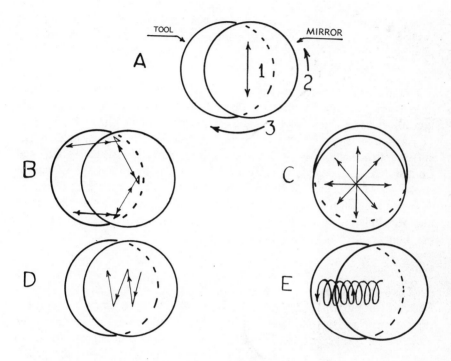

The mirror maker's strokes.

A. The fundamental motions:

(1) The movement of the mirror back and forth across the tool.

(2) The turning of the mirror at the end of a sequence of strokes.

(3) The movement of the worker around the work stand.

All other strokes are only variations of this one.

B. The chordal stroke.

C. Diametral stroke.

D. The "W" stroke.

E. The elliptical stroke.

In every case the arrowed lines represent the path taken by the center of the mirror.

The Diametral Stroke

Here the center of the mirror moves back and forth along *diameters* of the tool. Varying the length of this stroke changes the grinding effect.

Long strokes produce greatest abrasion at the center; short ones shift the action to the edges. If, during any stage of grinding, you find the curve is deeper at the center than it should be, you can cure the condition by shifting to a short stroke. The diametral stroke is a "safe" one; you will never get into serious trouble using it. But if used to the exclusion of all others, it may produce an irregular surface, or what mirror makers call zones, that is, high or low spots occurring in definite areas on the mirror.

The "W" Stroke

The name of this stroke is descriptive of its nature. From an overhang on one side, the mirror moves back and forth on chords, changing direction for each one, until it overhangs the same amount on the other side. The action of the "W" stroke blends irregularities in the surface of the mirror into a smooth curve since each irregularity is constantly attacked from different angles. It is more drastic in its action than the diametral stroke—a $\frac{1}{3}$ diameter "W" stroke has the same hollowing effect as a $\frac{1}{2}$ diameter diametral stroke. Because of its blending action, it is the stroke that we shall use in almost all stages of grinding and polishing after the preliminary stage of using chordal strokes.

Curved Strokes

These, like the "W" stroke, are used because of their blending action. The most popular is an elliptical stroke in which the center of the mirror traces larger or smaller ovals on the face of the tool. There are other varieties of curved strokes—circles, epicycles, etc.—but all are tricky to learn and use. Beginners should avoid them, especially during the grinding operation. Later on, when the mirror is to be polished, the elliptical stroke may help get rid of some irregularity which resists all other strokes.

THE EFFECT OF STROKES

In general, to repeat, long strokes deepen the center of the mirror and decrease the radius of curvature. Short strokes have the reverse effect. In between the two lies the "ideal" stroke which produces and maintains a spherical surface on the mirror. It is the $\frac{1}{3}$-diameter stroke. For a 6-inch mirror this will be 2 inches, that is, the edge of the mirror overhangs the tool for a distance of 1 inch at the end of each stroke. For an 8-inch it is $2\frac{2}{3}$ inches long for diametral and $2\frac{1}{2}$ inches for "W" strokes. This does not mean that every stroke taken must be an exact length, but it is important that the *average* come close to it.

If the curve becomes too deep in the mirror, it may be made shallower by using short strokes or by reversing mirror and tool and grinding with the tool on top. The reason for the latter is fairly obvious; if we grind a curve *into* the mirror when the mirror is uppermost, we also should grind the curve *out* of the mirror when the tool is on top.

STARTING TO GRIND

Your purpose at the beginning is to remove glass as rapidly as possible, so you will use the chordal stroke to "hog" out a hole in the center of the mirror. Here is the way to start:

1. Add abrasive and water; push the mirror back and forth on the tool so the center of the mirror is moving along a chord within an inch of the edge of the tool. Use a stroke as long as possible without the mirror's tipping off the tool at the end of the stroke. This distance will be about 4 inches for an 8-inch mirror. Use pressure on the mirror with each stroke, distributing it so most of the force is exerted on the *center* of the mirror and *edge* of the tool. Don't try to work too fast; about one back-and-forth stroke per second is enough.

2. About every six strokes, take a step to the left around the barrel so that the mirror moves back and forth along a new chord of the tool. Actually, it makes no difference whether you step to the left or the right, as long as you move your position far enough to produce strokes along a new chord. In whatever direction you start moving, keep on going until you have moved all the way round the grinding stand.

3. With each step to the left, turn the mirror slightly in the opposite direction (*counterclockwise*) so that the mirror will travel back and forth along a different mirror diameter. Again, it makes little difference which way you turn the mirror. Your object is to present a new diameter of the mirror to the tool. If you step to the left, turn the mirror to the left also, and vice versa.

These three motions should not be slavishly adhered to, since variety is the spice of life in mirror making. This doesn't mean that the beginner should attempt to invent new strokes and techniques. It does mean that slight variations in the motions mentioned above are good. Vary the length of the stroke a little. Vary the distance you step to the left; vary the amount you turn the mirror. If you fall into too much of a routine, you will grind the mirror unevenly. By changing the pattern of the three fundamental motions, mistakes made at one time will be corrected at another. As a matter of fact, this is why we can grind mirrors so perfectly. If we work according to a procedure that is fundamentally correct, the law of averages is on our side and we automatically compensate for errors which creep into our work.

The effect of these three motions is to hollow out the center of the mirror rapidly. At the beginning, most of the abrasive will be pushed over the sides of the tool, but once we have made a depression in the center of the mirror, more and more grit will remain in place and the work will progress more rapidly. Don't be afraid to use Carbo; dealers supply more than you need. As soon as the loud grinding noise ceases—you can also *feel* that the Carbo is cutting less effectively—lift off the mirror and add more grit and water. Mirror makers refer to the grinding down of each new charge of grit and water as a "wet." After a half dozen wets, wash the

Too much abrasive soon turns into mud, as shown here.

mirror and tool and inspect them. The tool will be ground more at the edges than the center, and the mirror will show the opposite condition. If you have been doing the work correctly, the edges of the mirror may not even be scratched, while its center will have a definite abraded area; and if you lay a ruler across it you can actually see the hollow beginning to form under the ruler.

Return the tool to its position, add more abrasive and water, and grind for another half dozen wets. Now you can actually measure the depth of the excavation in the mirror. Find several objects of different diameter— wires, pins, nails, etc.—and slide each one under the ruler until you find one that just fits. Measure the thickness of the object with a micrometer caliper: this is the depth you have reached. Better than this, however, is to use a template.

THE TEMPLATE

As we indicated earlier, the template is made by scratching a curve on a piece of cardboard or thin metal; $\frac{1}{32}$-inch aluminum plate, the kind you get in "do it yourself kits," is excellent for templates. The radius of the curve is equal to the radius of curvature of the mirror. When rough-grinding is complete, the curve in the template exactly fits the curve in the mirror all the way across any given diameter of the mirror. Much has been written about making *exact* templates, and many valuable amateur hours have been spent in producing one whose curve is perfect. But we are working to an approximate curve now and our template need not be one of finished perfection as long as it is reasonably accurate. So let's stop for a moment to consider the problem of making one equal to the purpose required.

Find a piece of board about eleven feet long. Drive a nail into it near one end. Make a loop in a piece of picture wire (any nonstretchable material will do). Stretch the wire tight along the board and measure a distance equal to the desired radius of curvature. In this case, since we wish a focal length of 56 inches, which requires a radius of curvature of 112 inches, we measure 112 inches along the wire and make a pencil mark on the board at this point. We tack a piece of heavy cardboard, such as the back of an 8×11-inch writing pad, to the board so that its long dimension is crosswise to the board and its center is over the mark. Then we stretch the wire tight again and, holding a sharp pencil or scriber at the 112-inch mark on the wire, strike an arc across the cardboard. Remove the cardboard and cut carefully along the line with sharp scissors. If we cut carefully enough, there will be no discernible bumps or hollows along the curve, and we have a workable template. Of course, if we wish a more permanent one, we can use metal or even thin glass and cut along the line with tin snips or a glass cutter. However, since a template is a one-purpose tool, we need not worry about permanency since we can always make a new one.

USING THE TEMPLATE

Test the hollow in the mirror's center with the template. If you can rock the template back and forth on its curve, the hollow is not yet deep enough. Continue grinding until it has become nearly deep enough to fit a section of the template. By this time the hollow will have spread out until it comprises about $\frac{1}{3}$ the diameter of the mirror. Now start grinding on chords of the tool which are closer to its center, and change the length of the stroke until about $\frac{1}{3}$ of the mirror overhangs at each end. The full stroke will not be over 5 inches, at this point, for an 8-inch mirror. As you use this stroke the central depression should not deepen but only become wider. As its diameter increases, grind over the chords closer and closer to the center of the tool, until finally the mirror's center is passing back and forth across the center of the tool and the curve extends all the way to the edge.

If you have been fortunate, the template now fits the full curve of the mirror, but you will probably find that the center of the mirror has a deeper curve than the areas near the edge. This is true if, when you place the template on the mirror and look under it toward a light, more light can be seen at the center than at the edges. Correct this condition by resuming grinding with center-over-center strokes of ⅓-diameter length until you have a spherical surface all over the mirror, that is, until the curve on the mirror matches that on the template. If the whole curve is deeper than that of the template, correct this condition by reversing positions of mirror and tool and grinding center-over-center strokes with the tool on top, using short strokes. This flattens the curve back to the curvature we wish.

A true spherical surface can be recognized by observing the action of bubbles between mirror and tool during the grinding operation. If, during a wet, a large bubble appears in the center of the mirror and retains its position there, the curve is too deep there. But if bubbles of the same size, after a few strokes, appear evenly distributed all over the grinding area, the two pieces of glass are in good contact throughout. Since there are only two possible surfaces which permit free motion of one over the other, the plane and the sphere, the mirror at this point must have a spherical surface. Another test you can use here is what is called the pencil test. Dry both mirror and tool. Then draw a series of light pencil lines across various diameters of the mirror. Place the mirror on the tool and make a half dozen short strokes. If the lines disappear at the edges but not the center you still have a central depression and you must continue grinding with the abrasive until it disappears.

The length of this description would make it appear that rough grinding must be a very long process. Actually, the whole operation can be completed in three hours without any strain on the part of the operator. But do not be surprised if your first mirror takes twice this long. If you end up with a smooth curve of the proper depth, the time will have been well spent.

TESTING FOR DEPTH OF CURVE WITH LIGHT RAYS

To make sure the curve satisfies the criteria above, we resort to a more accurate testing arrangement than the template. We make use of a very simple apparatus which utilizes light rays instead of mechanical measurements.

First we must refine the surface of the mirror to make it more reflective. As a preliminary, wash mirror, tool, the top of the grinding stand, cleats, and everything else that may have come in contact with the Carbo, including the hands and under the fingernails (a favorite hiding place for abrasive, where it remains unsuspected until the time comes for it to drop out and scratch the mirror). We clean up all cracks in the grinding surface and around the cleats with an old toothbrush and sluice everything away with water. This is where the perforated top of the grinding barrel is really appreciated.

Now replace the tool and resume grinding with No. 400 grit instead of No. 80 for two or three wets. The mirror's surface becomes smooth under this treatment and will retain a film of water to make it reflective. Screw a 7½-watt bulb into the socket of an extension cord. Place the mirror in the testing stand. Hold the light up near the right ear and adjust your position, facing the mirror, until you can see the image of the light bulb. Standing thus, bob the head up and down. If the reflection of the bulb in the mirror moves the *same* way as the motion of the head, you are inside the radius of curvature; if it moves in the *opposite* direction, you are outside the radius of curvature. Move away from, or closer to, the mirror until you reach a point where the reflection does not appear to move when you bob your head. At

Testing for radius of curvature. If image and light move in the same direction, the light is inside the radius of curvature. If in the opposite direction, the light is outside. If there is no motion of the image, the light is at the center of curvature.

this point you are very close to the radius of curvature of the mirror (or your ear is), and you should be able to determine the distance to within an inch. Take several measurements to make sure; and, if these measurements agree closely, take their average as the true distance. Divide this result by 2 to get the focal length. If it is several inches too long, you may wish to go back to No. 80 Carbo and, using straight ⅓-diameter center-over-center strokes, shorten it to the desired length. If too short, the best plan is to accept it and go on to fine grinding. After all, the only reason for working to an exact focal length is possibly one of pride in accomplishing the feat. A difference of two or three inches in focal length means only that you will have to adjust the length of the telescope tubing to accommodate it.

Wet paper toweling holds the mirror in place as effectively as cleats during grinding operations. Note the position of the hands on the mirror.

The mirror must, however, be spherical before you go on to fine grinding. You have already made a rough test for sphericity by watching the bubble patterns (grinding with No. 400 grit produced a fine opportunity to make this test). Now you can apply another test which will give you a little more exact information.

Wash and dry the mirror carefully, then hold it flat in the hands, ground surface up, and catch the reflection of an overhead bulb in it. The test is more accurate if the bulb is unfrosted and the reflection of the filament itself can be seen. You will have to experiment a little with the angle at which the light strikes the mirror (the flatter the angle at which the light is seen, the brighter the image will be), but you will eventually find a spot at which, when the filament reflection is in the center of the mirror, you will see a good image. Tip the mirror one way or another so the image moves from the center toward an edge. If there is a smooth change in the intensity of the light as the image moves from center to edge, the mirror is spherical. But if there are abrupt changes in the intensity, the mirror has an uneven surface. Brighter images mean high spots on the surface, dimmer ones indicate low spots. Do this test over all diameters of the mirror. Slight changes are permissible since they will be ironed out by succeeding fine grinding, but marked changes in the quality of the reflection mean that we must return to rough grinding and bring the mirror to sphericity before proceeding further.

But let us assume that you have succeeded in producing a spherical mirror of a focal length that satisfies you. You are now ready to go on to the next step, fine grinding or smoothing the curve that you have produced.

The Mirror: Fine Grinding

Most of the operations you undertake during the fine grinding stages consist of repair work. You have ground a curve into the mirror, but in so doing you have ruined its surface from an optical viewpoint. The glass is now covered with pits, scratches, and other irregularities which must be removed before you can polish it. These can be ground away with No. 120, but they will be replaced with smaller ones from the No. 120 itself. You remove these in turn with the next grade of abrasive, and so on down the line until the final irregularities are of such a small size that they can be removed with polishing agents, working on a pitch lap.

In order to perform these operations with an understanding of what you are doing, rather than mechanically following directions, it is helpful to know a little more about abrasives and how they work. Any material harder than glass can be used as an abrasive, depending on the way it is used. If it is placed between two surfaces of approximately equal hardness, it scours and abrades each of the surfaces as it rolls between them. But if it is embedded in a soft material such as wood or pitch, it produces a shaving action which tends to polish the glass surface being worked upon it. This shaving action is a very limited one, however, and removes only minute quantities of glass. If the surface is heavily abraded, a polish will be produced but only between the deep pits and scratches. This is the reason we reduce the size of the pits by using progressively smaller abrasive particles, instead of starting to polish immediately.

Manufacturers grade their abrasives carefully, but are not always successful in doing so completely, especially in the larger sizes. No. 120

Carborundum, for example, is almost certain to contain some particles the size of No. 80. But they are relatively few, so that when we add a charge of No. 120 to the tool, there will be only a few grains of No. 80 sticking up here and there. When we place the mirror on the mixture, the large particles produce scratches and pits before the pressure of the mirror breaks them down to the size of the rest of the particles. At this point they work with the smaller abrasive in removing material around the deep pits, and the result is that all pits become shallower. Since there will always be large-sized particles in each new charge of abrasive, it would seem hopeless that we could ever eliminate all pits, because each new charge would produce new ones.

There is a way out of the difficulty, fortunately. When we add a new charge to the tool, we place the mirror on it and, without applying any pressure at all, gently move the mirror back and forth for several very short strokes. This breaks down the large particles without digging them into the glass; the pits they produce in this way are much smaller than they ordinarily would be. As we go from abrasive to abrasive, repeating this process each time we start a wet, pits and scratches will be reduced to a minimum.

The difference in grain size of abrasives and the corresponding action each has on glass become apparent if we study the table below for a moment.

Mesh Size (used by most companies in grading abrasives)	Size in Inches	Size in Microns (1 micron = .001 mm = .00004 inch)	Elutriative Time	Other Designations
60	.0117	293		M60
80	.0075	185		
100	.0059	148		M100
120	.0049	123		
180	.0034	86	1 min.	M180
220	.0029	73		
280	.00156	39		
400	.00135	34		
600	.00100	25		
700	.00088	22	5 min.	M302
850	.00068	17	10 min.	M302½
1000	.00062	15.5	20 min.	M303
1300	.00044	11	40 min.	M303½
1800	.00032	8	60 min.	M304
3000	.00020	5		M305

Bausch and Lomb uses mesh sizes to grade their abrasives, while the American Optical Company uses the designations shown in the far-right column. If each designation is translated into size in microns, much confusion can be avoided. Thus it can be seen that mesh size 3000 corresponds to M305, since each contains particles 5 microns in diameter. Several suppliers use micron size to identify their abrasives. Garnet grinding powders are ordinarily designated this way. No. 8 garnet, for example, is a fine garnet abrasive whose particles are 8 microns in diameter and corresponds to No. 1800 and M304 in the table above.

The column in the table marked "elutriative time" refers to a method of grading abrasives used by manufacturers and amateurs alike. If powdered abrasive is added to a column of water 3 feet high, to which a deflocculant (such as a detergent) has been added, the large, heavy particles sink to the bottom immediately. The lighter (and therefore finer) particles remain in suspension for a time whose length is in inverse proportion to their size. If, after a certain period of time, the water containing the suspended material is siphoned off and put aside until all the suspended material has finally settled, the resulting layer at the bottom will contain no particle over a certain size. Thus "5-minute emery" contains only particles of a size so fine they will not settle out of suspension in 5 minutes. The table shows that such particles are the same size as No. 700 mesh or M302, 22 microns. Ordinarily it is not necessary to refine grades of abrasive in this way, but if scratches appear on a mirror which cannot be explained in any other way, it is a good idea to adopt this method of checking the finer grades, since they have probably become contaminated.

THE FINE-GRINDING PROCESS

Make sure that you have cleaned up thoroughly from rough grinding. Sprinkle the tool with No. 120 grit, spread it around with the finger tips, and wet it from the spray bottle. Place the mirror on it and take a few short strokes very slowly and gently. The strokes from this time on will be $\frac{1}{3}$ "W" strokes, although they should actually be a little shorter than $\frac{1}{3}$ the diameter: $2\frac{1}{2}$ inches instead of $2\frac{2}{3}$ inches. But we add a variation to the stroke which we shall use to the end of fine grinding. In addition to the straight back-and-forth motion, the mirror gradually moves across the tool from a 1-inch overhang on the right to an equal amount on the left. Continue to apply pressure, although there is no longer need to press down as hard as you did in rough grinding since you are now attempting to smooth the curve, not deepen it. The remaining elements of the stroke are the same as those for rough grinding. The mirror is rotated slightly, counterclockwise, and the operator steps to his left about every six strokes. Again we emphasize that no devoted adherence be made to make the strokes exactly as described. Random variation of all elements of the stroke is the ideal you should strive for, hoping that the average will correspond to what you want to achieve.

Continue each wet until you can tell from sound and feel that the abrasive is no longer cutting. Don't try to prolong the wet to get the last bit of abrasive action from each charge. When the grinding "feel" changes from a sandy action to a muddy one, you have gone far enough. Old hands at mirror making agree that a small amount of abrasive, well wet down and changed often, is the best insurance for good grinding action. Wash the spent abrasive away with a few squirts of water from one of your plastic squeeze bottles, add fresh abrasive, and continue grinding. Each wet at this stage should last from 6 to 8 minutes. If the abrasive begins to dry out in the middle of the wet, slide the mirror part way off the tool and add some more water. After two wets, wash both mirror and tool, then reverse their positions on the grinding stand. Grind through two wets with the tool on top, using slightly shorter strokes than when the mirror is on top. This is to prevent grinding the edge of the mirror too much. Grinding with the mirror on top tends to shorten the focal length of the mirror; when the tool is on top the reverse is true. If the positions are alternated regularly, the depth of curve will remain the same.

Grinding with the tool on top to lengthen radius of curvature. The serrations are typical of this kind of ceramic tool.

How long should we continue with one abrasive before going on to the next? One way to find out is to grind for 15 minutes, then wash the mirror and inspect it under a strong light. Use a magnifier or an eyepiece if you wish, but usually your eyesight is good enough for this test. Look for the following:

1. Is the appearance of the mirror uniform from edge to edge? The center should not be smoother than the edges, or vice versa.
2. Are the pits the same *average* size? Don't worry about how large or how small they are, as long as they all look alike.

You won't be able to answer "yes" to either of these questions after only 15 minutes of grinding on any particular abrasive size, but you may well do so after an hour. When this happens it is time to go on to the next size. To be safe, let's accept one hour as the minimum grinding time for any one abrasive.

Another examination of the table of abrasives given previously (page 41) helps us in this decision. No. 120 abrasive is made up of particles about ¾ the size of those of No. 80; but the next size, No. 220, is only ⅜ of No. 80. If we do not eliminate the pits left from No. 80 with No. 120, the chances of doing so with No. 220 are small indeed. Some mirror makers deliberately scratch a mirror with a diamond after the completion of rough grinding, attempting to make the scratch deeper than any pit left behind by the coarse abrasive. When this scratch grinds out, they feel sure that the deep pits have gone with it. Pits are tenacious in that deep ones hide among the more shallow ones and don't become visible until the completion of fine grinding or the first stages of polishing. When they appear at this stage, there is nothing to do except go back to the coarser grades of abrasive to get rid of them, which means that all subsequent stages have to be gone through again. Consequently, spending longer than is really necessary on No. 120 and its successor, No. 220, is an insurance against being forced to repeat work already done.

After grinding for an hour or so—and this is actual grinding time, not total time consumed, since we can't count the time spent in adding abrasives, inspecting the mirror, washing operations, etc.—we wash everything again as thoroughly as we did after the rough grinding was finished, and go on to No. 220. During the washing-up period, take a few moments to inspect the bevel on the mirror and tool. If it needs attention, now is the time to do it, since the bad effects of chipping become more serious as the mirror is more finely ground. The bevel should be ⅛ inch wide at this point. If it isn't, widen it. Check the focal length carefully, using the small light bulb as a testing device. This testing will become easier as the surface of the mirror is more refined since the smoother surface makes it more reflective. Check for sphericity, using an overhead light, as described on page 39. Be sure the brightness of the image dims gradually and smoothly as it moves toward the edge of the mirror. If you have been careful with the No. 120 stage, you have nothing to fear as to the curve on the mirror, but it is always wise to check, just in case.

The No. 220 grinds much more smoothly and easily than its predecessor. You will find that you need to do a little experimenting to discover the proper amount af abrasive and water you need for each wet. As in previous grades, you can tell by noise and feel when the abrasive is really cutting. If the abrasive doesn't seem to spread out evenly over the tool at the beginning of a wet, add a little detergent to the water (Dreft, Glim, or any of the common dish-washing detergents are good for this purpose). Make this a weak solution; 2 to 3 per cent is strong enough. Otherwise you will be up to your ears in soapsuds. A detergent works in two ways. It acts as a deflocculant, which means that it helps to prevent the abrasive from gathering into small bunches, and it is a wetting agent, which makes the water spread evenly over the surfaces of tool and mirror. Continue alternating position of tool and mirror every two wets or, if you are willing to spend the extra time washing up, every wet.

After 15 minutes, wash and dry the mirror and inspect it for pits. This time use a strong transmitted, or diffused, light, along with a magnifier or eyepiece so that you can make a close examination of the surface. At first the surface may appear perfectly uniform, but as you look more closely you will see a scattering of points of light against a smoother background. Each of these represents a pit, and they should be scattered evenly from the center to the edge of the mirror. Mark the position of the largest ones you can find by using a china pencil on the back of the mirror. When these grind out, you may be sure that other large pits have gone with them and that all those remaining are of a uniform size, but continue to grind for the remaining 45 minutes anyway.

You have now spent at least 60 minutes with each of the first two grades of fine grinding abrasive. The remaining grades will require 60 minutes or less for each. The table below gives a safe margin of time per abrasive. If you follow it closely you will end up with a fine-ground mirror, free from pits and, if you have been careful, from scratches and other defects.

ABRASIVE	TIME
No. 80	3 hours, or until mirror is spherical and of the proper depth.
No. 120	60+ minutes (10 wets)
No. 220	60+ minutes (10 wets)
No. 280	60 minutes (8 wets)
No. 400	60 minutes (8 wets)
No. 600	60 minutes (8 wets)
No. 900	40 minutes (5 wets)
M305	40 minutes (5 wets)

The time given in the table for each abrasive is an *average* one. Slavish adherence to any table such as this is apt to waste time. Check for pits carefully; when you are satisfied they are all the same size, go on to the next abrasive. Sometimes you can beat this time schedule, sometimes you may run over it.

Grinding with No. 280 will follow the same pattern as with No. 220 except that you will probably find that you need to stop more often in the middle of a wet to add water. This stage should give you little trouble, since you are now thoroughly accustomed to the strokes used and to the general behavior of mirror and tool. You will find that you have a tendency to rush through this stage, but you should resist the inclination and be as careful and painstaking as you have been in previous stages.

In using No. 400 and subsequent finer grades of abrasive, you may apply the abrasives to the tool directly from their shakers or mixed with water and applied as a suspension. If the latter course is adopted, place the abrasive in a plastic squeeze bottle, add water, and shake. When the bottle is shaken, it must show a solid color of the abrasive. If it appears thin and watery, you have too thin a mixture.

Allow the mixture to settle for a few seconds, then apply a few drops to the center of the tool. Add a pinch of Dreft or a drop of liquid detergent and spread it with the finger tips. The total amount of abrasive to be added for each wet is about the size of a pea, so you will have to estimate this if you are adding it from a suspension.

Place the mirror carefully on the tool and move it slightly with the ear close to the mirror. If no crunching or grinding is heard, it is safe to go ahead with the wet. But if the sound indicates that there are sizable particles of abrasive present, *lift* (don't slide, since this will make scratching a certainty) the mirror off, wash everything, and try a fresh charge.

As with No. 280 and previous grades, grinding with No. 400 should proceed with mirror and tool changing positions on alternate wets, or at the end of every second wet. The surfaces of the two pieces of glass are now much closer together, the grains of abrasive are smaller, and the noise from the operation sounds more like the rustle of leaves than an actual grinding sound. At the beginning of a wet, the mirror will appear translucent, but an opaque film will soon appear as the abrasive becomes ground down. The whole process appears easier, and the operator, having gotten this far, is inclined to proceed with much confidence. But this is a danger point, for it is at this stage that scratches are most likely to appear on the mirror. They may appear in several ways:

1. Contaminated abrasive. If precautions are taken as outlined above, there is little danger of this.
2. Chipped edges and a resulting small piece of glass getting between the surfaces.

3. Trying to wear the abrasive down too far before adding a fresh charge, or by allowing it to become too dry toward the middle or end of a wet. A lump of abrasive, even if made up of small pieces adhering, acts very much like a large solid chunk and will produce a scratch.

4. Seizure of the disks. This usually is the result of too little abrasive and water. The disks lock themselves together and become extremely difficult to separate without scratching. If this happens, there are several ways of separating them. The least risky is to stand the locked disks on edge under a stream of water from a faucet. If at the end of a few minutes, they are still locked, more rugged means (and also more risky) must be taken. Borrow a carpenter's wooden clamp and apply it across the projecting edges of mirror and tool. Tighten up the clamp hard, and immerse the whole arrangement in a bucket of water, at the bottom of which you have placed a piece of old blanket or other soft material. Failure to separate them by this means leads to the last and most dangerous expedient. Place the projecting edge of the mirror against a wooden stop nailed to the top of the table. Now place a curved end of a 2×4 against the tool and strike the other end a sharp blow with a hammer. Hold your breath when you do this; it has been known to work without breaking either of the two pieces of glass. Most likely it will break a large chip from the edge of the tool but this is not as serious as it sounds, for the fine grinding can continue even with a defective tool.

Having observed all these precautions, continue grinding with No. 400 for the required time (60 minutes).

THE FINAL STAGE OF FINE GRINDING

You now have three abrasives yet to be undertaken, No. 600, No. 900, and M305. You must spend an hour on No. 600 and about 40 minutes each on the other two.

You must be especially careful with these three stages, for it is here that the danger of scratching is greatest. The likelihood that the disks may stick together is also greater; hence for all three stages wash the mirror after every wet. The abrasives should be applied from suspension, and the use of detergent with each wet will be more and more helpful as the size of the abrasive becomes smaller. Especial care should be given the first few strokes of each wet; actually try to hold the mirror up a little during these strokes. When you are sure that there is no danger of scratching from possible contamination of the abrasive, complete the wet with little or no pressure other than the weight of the mirror itself.

Alternate the positions of tool and mirror with every wet throughout the final three abrasives.

When you have completed work with the No. 600, wash the mirror thoroughly and inspect it for pits, holding it against a bright light and using a magnifier. The surface should present only a pattern of the small

Using a light and an eyepiece to check for pits.

abrasive pits left from the No. 600 itself. Since No. 600 and its successors are very tenacious materials because of their small size, you may mistake a particle of the abrasive itself for a pit. Stroke the suspected pit with the ball of the thumb. If it moves, it is not a pit. Check for sphericity and focal length.

You may experience difficulty in making good contact between mirror and tool with the final two abrasives. The tiny bubbles which have appeared at the beginning of each wet, and which we have disregarded until now, tend to hold the glass surfaces apart. This can be overcome, after the usual tentative preliminary strokes, by slowly rotating the mirror while pushing it across the tool. The bubbles will work themselves over the edge, and will leave a smooth film of water and abrasive behind. Now go ahead with regular strokes. Watch out for sticking in the middle of a wet. You are using only a very small amount of water now, and it will evaporate very rapidly. Don't leave the disks together if you are interrupted in your work. Sometimes only a few seconds is necessary for them to become stuck if they are not moving relative to each other.

The final wet of M305 should be prolonged for twice the ordinary period, with water added at frequent intervals. M305 (see table on page 41) is fine emery. It is especially useful for the final stages of grinding because the pits it leaves behind are more shallow than those of other abrasives. Shallow, wide pits polish out much more rapidly than deep, narrow ones. During this time the mirror should be on top and should be handled carefully and cautiously. At the end of the wet, wash and dry the mirror with twice the usual thoroughness. Wipe it with a soft cloth (an old towel that has been through the laundry many times is excellent for this), and wipe off any remaining lint with the under part of the forearm or the heel of the hand.

The fine ground mirror has a velvety feel and needs only a suggestion of dampness on its surface to become transparent. When dry, it passes enough light so that it is possible to read 10- or 11-point type (the size of this type) through it at a distance of several inches. It will have a spherical surface, free from pits and scratches. If you have achieved such a mirror, you are well on your way to possessing a very fine optical surface, one that will respond readily to the final steps of polishing and figuring.

In this intermediate stage of fine grinding, the mirror is only partly transparent when dry. But at the end of fine-grinding operations, the type at the edge of the mirror should be clearly legible.

CHAPTER 5

The Principle of the Foucault Test

T HE FINE-GROUND MIRROR is ready to be polished, but it will have to wait while you complete two very important intermediate steps. One of these is to make a Foucault tester, by which you can make accurate tests on the surface condition of the mirror, and the other is a pitch lap to polish it on.

Let's talk about the Foucault tester first, since you can use it to get a better idea of your fine-ground mirror even before you polish it. The great physicist-astronomer Jean Foucault devised the test almost a hundred years ago to replace the laborious and time-consuming tests, using stars, which all mirror makers had used. Prior to this ingenious test— all the more remarkable because it is so simple—making and testing astronomical mirrors was more of an art than a science.

Because we shall use the Foucault test from now on, it is important that we understand the principle as well as make the apparatus by which we actually apply the test to the mirror. Here is the way it works.

If a very small point of light is placed at the center of curvature of a mirror (the point at which the radii of curvature converge), its diverging rays will strike all points of the mirror's surface. If the mirror is perfectly spherical in shape, each ray striking the surface will be reflected back to its source. Now suppose the point of light is moved slightly to the right but still remains at a distance equal to the radius of curvature. In this case the reflection will take place as before except that the point of intersection of the reflected rays is shifted to the left. In other words, a cone of light is reflected from the mirror which comes to a focus to the left of its original source. The exact location of this point can be found by placing a piece of ground glass in the area. After the position of the point has been determined, the ground glass is removed and the eye is placed in this position. The mirror will appear to be fully illuminated, since every point on its surface is reflecting a ray which originally came from the light source.

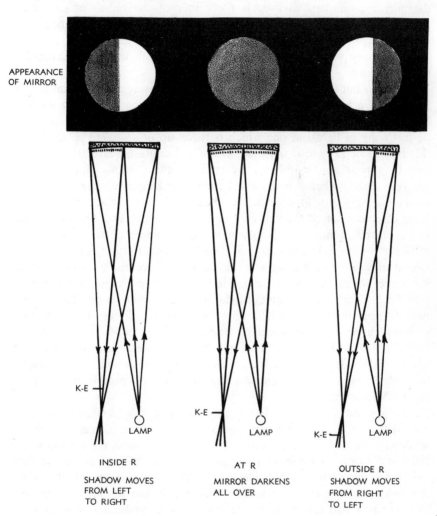

APPEARANCE
OF MIRROR

INSIDE R

SHADOW MOVES
FROM LEFT
TO RIGHT

AT R

MIRROR DARKENS
ALL OVER

OUTSIDE R

SHADOW MOVES
FROM RIGHT
TO LEFT

Using the Foucault test to find the radius
of curvature (R) of a spherical mirror.

Suppose now we cut part of this reflected cone of light with a razor blade or other sharp edge. If we cut the cone in front of the place where it comes to a focus, a shadow will appear to cross the mirror, moving in the same direction as the blade. If we cut the cone at its focus (or the point where the reflected rays cross each other), the mirror will darken instantaneously all over its surface. But if we cut the cone after the light rays have crossed each other, or at a point in back of the focus, a shadow will cross the mirror in the opposite direction to the motion of the razor blade. The point where the motion of the razor blade causes the mirror to darken all over is the center of the radius of curvature, provided that razor blade and light source are the same distance from the mirror. If they are not, we take the average distance of the two. This is a very sensitive test for focal length, which will be half the distance we have just measured.

The Foucault test goes far beyond a simple measurement of focal length, however. We have assumed in our discussion that the mirror is spherical, which will rarely be the case. If it is not spherical, it will not reflect a perfect cone of light but one which is made up of scattered rays. Cutting this bundle of rays with the razor blade will cause a varied shadow pattern to appear on the mirror. By studying the shadow pattern we can see very clearly where bumps and hollows are located on the mirror and, because a very small imperfection on the mirror's surface will cause a large displacement of the reflected ray, these bumps and hollows appear to be magnified many times. As a matter of interest, the magnification of imperfections is in the order of 100,000 times. Furthermore, a distinctive shadow pattern will appear on the mirror according to the particular curve which has been ground into its surface. As we have said, a spherical surface causes an even darkening all over, a paraboloid produces a doughnutlike shadow, and so on. We will discuss these shadows and the curves they represent at length when we come to the chapters on testing and figuring the mirror. For the present, however, we shall use the Foucault tester only for an accurate determination of focal length, as a check on sphericity, and for the detection of imperfections on the surface of the mirror.

The apparatus in its simplest form consists of a light source—an electric light bulb inside a tin can, one side of which is pierced to provide a pinpoint or slit of light—and a knife edge mounted on a slide which has freedom of motion in two directions—toward the mirror and at right angles to the beam of light. There are almost as many ways of making the apparatus as there are amateurs who use it, but we shall confine our discussion to two forms: a simple one, easy to make and use, and another for those who like the highest possible precision and who are willing to take the time and trouble to make it.

THE SIMPLIFIED FOUCAULT TESTER

Get a 7½-watt, 115 volt AC night light, the type which plugs into a baseboard socket. Ask for a frosted bulb instead of a clear one. This unit will be placed in a $4 \times 1\frac{1}{2} \times 1\frac{1}{2}$-inch can, the kind that baby powder is sold in. Pry off the part that twists to release the powder (the top section with the holes in it) and throw it away. Then remove the top of the can and screw it, open end up, to a $\frac{1}{2} \times 12 \times 15$-inch plywood board, as shown in the photograph.

With a pair of tin snips, cut a slit in the side of the can to accommodate the plug of the light unit. Cut a $\frac{3}{8}$-inch round or square hole in the side of the can opposite the center of the light bulb. Just below this hole, attach a thin piece of wood with screws or rivets, placing the rivets low enough so that two single-edged razor blades may be inserted under the wood to form a vertical slit across the $\frac{3}{8}$-inch hole as their edges approach one another. Now mount the light unit in the can.

Drill a $\frac{3}{8}$-inch hole in the back of the can so that a 2-inch section of $\frac{3}{8}$-inch dowel can be inserted to turn the light off and on. These lights ordinarily have a rotary switch, turned off and on by a cylindrical knurled knob. If so, drill out the end of the dowel so it fits the knob snugly. If the switch handle is the wafer type, slot the end of the dowel to fit. Now place the can and light unit in its socket (the top of the can) on the plywood board.

Simple Foucault tester, rear view. The white
scale at lower left is part of the Barr scale.

We spoke of a *point* of light when we discussed the principle of the Foucault test, but we have produced a *slit* of light in the actual construction. For practical purposes the slit is as good as if we had used a pinhole in the side of the can and is much easier to use. If the knife edge is adjusted parallel to the slit and then moved into the reflected beam of light, as it moves across the beam it will cut off the light as though it came from a point source. The slit works best if it is narrow, about 10–20 microns, (roughly the thickness of a piece of tissue paper). The only exacting requirement of the slit is that its edges be parallel to each other and to the knife edge.

The Knife-Edge Apparatus

It is in this part of the Foucault tester that we find the most variation in structure and precision. Actually, all we want is a device to cut the reflected cone of light from the mirror at right angles to its axis. But for reasons which appear later we shall also want to know exactly how far from the surface of the mirror the knife edge cuts the cone of light. So we shall have to construct a holder for the knife edge which can be moved forward or back relative to the mirror, and to make a device which will measure this motion to 100th of an inch. We also need to move the knife edge at right angles to the cone of light. But we do not have to measure the amplitude of this motion; and, since it is usually very small, pressure on one side of the testing stand is enough to produce it.

Simple Foucault tester, front view. The slit between the razor blades is exaggerated in this picture so that the light source can be seen.

We start by mounting the knife edge (a single-edged razor blade) on a block of wood (see photograph) which can be moved back or forth along a guide strip of wood. The guide strip is placed as close to the lamp unit as we can make it and is oriented parallel to the axis of the beam of light. We wish as little separation as possible between knife edge and slit, especially in testing short-focus mirrors, since this cuts down errors of astigmatism.

The Barr Scale

The problem of measuring very small back-or-forth motion of the knife edge can be solved by using what is called the Barr scale. To make such a scale, mark off a series of 6 parallel lines on a thin sheet of brass $2\frac{1}{2} \times 3\frac{1}{2}$ inches. These lines must be exactly $\frac{1}{10}$ inch apart, as shown in the diagram. Draw 6 more lines, perpendicular to the first set and parallel to each other, $\frac{1}{2}$ inch apart. You now have a rectangle $\frac{1}{2}$ inch long and $2\frac{1}{2}$ inches wide, divided into 25 smaller rectangles. Draw diagonal lines as shown. If possible, all lines should be drawn with a ruling machine. Lacking that, use as the next best thing an engineer's steel rule, divided into 100ths of an inch. These can be obtained in the 6-inch size from Brown and Sharpe for about $2.50. Use a sharp scriber or an engraving tool and check and recheck the measurements until you are sure they are accurate.

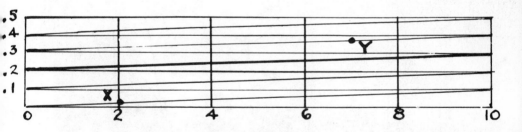

The Barr scale.

Since the distance between each of the main divisions on the Barr scale is $\frac{1}{10}$ inch, any point on the diagonal connecting two of the main divisions must be a fractional part of this distance. Since the whole length of the diagonal represents a vertical distance of $\frac{1}{10}$ inch, half of it must be $\frac{1}{20}$ inch, $\frac{1}{5}$ of it must be $\frac{1}{50}$ inch, and so on. For example, the point marked X represents $\frac{1}{5}$ of $\frac{1}{10}$ inch, or .020 inch, while the point marked Y we would estimate as .370 inch. Note that unless the diagonal actually intersects one of the vertical lines, the point must be estimated to the third significant figure, so point Y could be .369 or .371. However, we want figures accurate only to 100ths, so this will be close enough for our purposes.

The brass plate engraved with the Barr scale is mounted flush with the surface of the plywood. Now mount an indicator, a flat piece of metal with a thin, *straight* edge, on the bottom of the block which holds the knife edge. This indicator must be at least the width of the scale so it will cover the full width of the horizontal divisions, and its edge must also be parallel to them. Practice sliding the block to various positions of the scale and read the distances indicated. After a little practice, getting an accurate reading will become easy.

There are several concerns that manufacture Foucault testers, complete with micrometer adjustments in two directions. For those who would prefer to concentrate their energies on making mirrors and other parts of the telescope, buying or renting a tester will seem a good solution. One company will even lend you a tester if you purchase a mirror-making kit. Buying one costs between $25 and $40.

THE MACKINTOSH TESTER

It is possible to build a very precise tester yourself, especially if you have access to some simple metal-working apparatus. You will need a

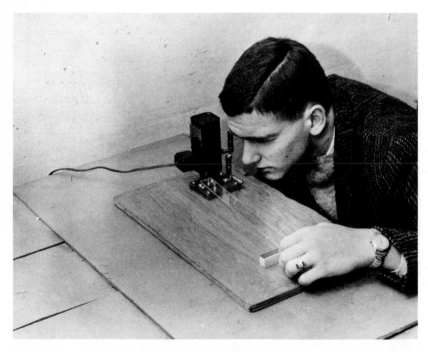

Using a precision Foucault tester. In this photo the camera cable
has been temporarily replaced by a thumb screw.

drill press and some taps and drills for machine screws, plus files, a hacksaw, and other simple hand tools. The design we give here was worked out by Allan Mackintosh of Glen Cove, New York, who adapted it from an original design by Kelvin Masson of St. Louis, Missouri. It is the best one we have seen and used and, if made carefully, gives very precise readings for knife-edge motion.

This tester is based upon the principle of the lever. The knife edge is mounted on a slide which is moved back and forth in a grooved track by means of a lever. The slide is attached to the short arm of the lever, and the long arm moves back and forth over an accurately ruled scale. A large movement of the scale arm produces only a small motion in the short, or slide, arm. In the Mackintosh tester the ratio of the two arms is 10 to 1. Thus a movement of 0.1 inch of the scale arm provides a knife-edge motion of only .01 inch. This is called the amplification factor of the device, and by reading the scale accurately we may measure knife-edge movement to within .001 inch.

Attached to the main slide and moving with it is a cross slide to provide freedom of motion. The cross slide moves at right angles to the main slide and is also provided with a grooved track. This is difficult to describe in words, but a glance at the diagram shows how these slides work on each other so that motion of the lever does not cause sticking or binding.

Materials Needed for Mackintosh Tester

Lever arm: Lucite or aluminum $14'' \times 1'' \times \frac{1}{8}''$
Main slide: ground steel stock $3'' \times 1\frac{1}{2}'' \times \frac{1}{8}''$
Channels for main slides: 4 pieces ground steel stock $3'' \times \frac{3}{4}'' \times \frac{1}{8}''$, 2 pieces ground steel stock $3'' \times \frac{1}{2}'' \times \frac{1}{8}''$
Cross slide: 1 piece ground steel stock $\frac{3}{4}'' \times \frac{1}{2}'' \times \frac{1}{8}''$
Channels for cross slide: 2 pieces ground steel stock $\frac{3}{4}'' \times \frac{1}{2}'' \times \frac{1}{8}''$
Base: plywood $10'' \times 18'' \times \frac{3}{4}''$
  3 pieces $\frac{1}{8}'' \times 1''$ commercial dowel pins
  3 pieces $\frac{1}{8}''$ drill-jig bushings
  1 6''-steel scale, Brown and Sharpe, divided in 100ths of an inch
  1 camera release cable, 24'' long
    aluminum stock, $\frac{1}{8}''$ sheet (for knife-edge support)
    razor blade (single edge)
All sections of the main and cross slides can be sawed from 2 pieces of ground stock 18 inches long, one of them $\frac{1}{8} \times 1\frac{1}{2}$ inches and the other $\frac{1}{8} \times \frac{1}{2}$ inch.

The main slide is a piece of stock $3 \times 1\frac{1}{2} \times \frac{1}{8}$ inches. It travels in two channels, each made out of a sandwich of a piece of $3 \times \frac{1}{2} \times \frac{1}{8}$ between two pieces of $3 \times \frac{3}{4} \times \frac{1}{8}$. The individual pieces which make up the channels are separated from one another by aluminum wrap shims to allow

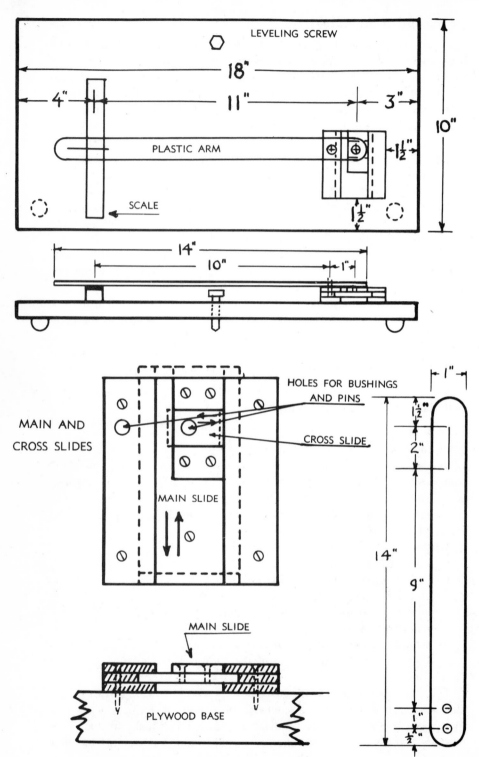

The Mackintosh adaptation of the Foucault test device.

clearance for the main slide. The channels are clamped together and drilled with a No. 32 drill and the top piece countersunk for a No. 4 wood screw. Then take them apart and redrill the center piece of each sandwich with a No. 28 drill. Place aluminum shims between the layers of each sandwich and assemble the whole arrangement on the plywood board $1\frac{1}{2}$ inches from each edge at a corner of the board. Since the holes in the center part of each channel are a little larger than those of the top and bottom pieces, these can be pressed in with the fingers while the apparatus is being screwed to the board. This will insure a smooth, easy motion to the main slide and will prevent side play in the channels. Grease the edges of the main slide, rather than use oil, because grease works better on this type of bearing surface and takes up less room than oil.

Drill a $\frac{1}{8}$-inch hole, $\frac{3}{4}$ inch from the end of the left-hand channel and gently hammer in one of the dowel pins. Be careful here or you will upset the assembly as a whole.

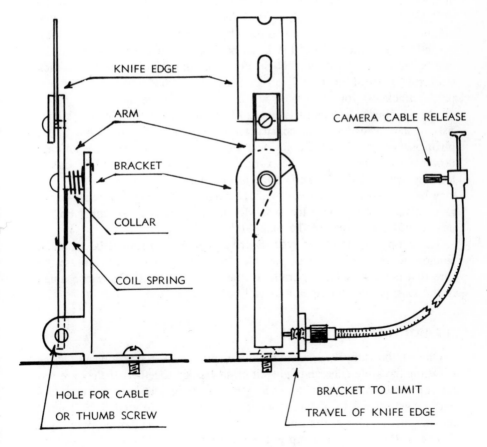

Knife-edge holder.

The cross slide consists of three pieces of $\frac{3}{4} \times \frac{1}{2} \times \frac{1}{8}$-inch stock assembled on the main slide as shown. Drill the end pieces first, then clamp all three pieces together after having placed tissue-paper shims between their edges. Place them on the main slide and mark where the drill holes come. Remove the main slide and drill and tap it for machine screws. Assemble all four pieces with shims in place, then force out the cross slide and remove the paper shims. When greased and reassembled, the cross slide should work as easily and cleanly as the main slide.

Mark the cross slide at a point exactly 1 inch from the center point of the dowel on the main slide, drill a $\frac{1}{8}$-inch hole, and force in the dowel pin, using the jaws of a vise to apply the pressure.

Round off the ends of the Lucite arm. At a point $\frac{1}{2}$ inch from one end, drill a hole whose size is equal to the outside diameter of the $\frac{1}{8}$ inch drill-jig bushing. Be careful here, since Lucite has a tendency to star or crack if subjected to much pressure. Drill another hole whose center is exactly 1 inch from the center of the first. Scribe a longitudinal line along the center of the arm. Fill in the scribe line with black. Force the drill jig bushings into the holes provided for them (careful, here!), then mount the Lucite arm on the two dowel pins on the main and cross slides.

Under the center of the black scribe line attach a strip of wood 10 inches from the pivot of the Lucite arm, and mount on it a 6-inch steel rule marked in 100ths of an inch. The wood strip must be the proper thickness for the Lucite arm to just touch the steel rule.

The knife-edge assembly is made of $\frac{1}{8}$-inch sheet aluminum. It consists of a bracket and arm constructed according to the diagram. These pivot on one another at a point near the top of the bracket. Use another dowel pin and bushing for the pivot. The cable release should be of the locking type to hold the knife edge in one position. The knife edge itself may be any flat piece of metal with a straight, sharp edge. A single-edged razor blade is very useful for this. Mount the knife-edge assembly on the near right-hand corner of the main slide.

The lamp may be of the type described earlier and should be mounted as close to the knife edge as possible. In lining up the tester for use, be sure that the knife edge is parallel to the slit in the lamp. Otherwise much of the sensitivity of the apparatus will be lost.

THE HOT WIRE LIGHT SOURCE

One of the difficulties encountered in making a good Foucault tester is getting the knife edge and light slit close enough together. If they are too far separated—more than an inch—various astigmatic effects are introduced. An ingenious device for bringing knife edge and light source infinitesimally close is the one originated by E. G. Onions of Bulawayo, Southern Rhodesia. The lamp is completely eliminated and is replaced by a fine, incandescent wire placed in front of the knife edge. In this setup the motion of the knife edge *exposes* part of the cone of light instead of cutting into it, as with more conventional devices.

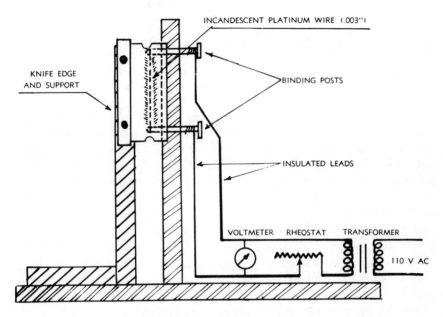

The hot wire light source.

Stretch a piece of .003-inch platinum wire (see appendix for sources and cost) across two binding posts in front of the knife edge as shown in the diagram and connect it to a 6-volt battery or 110-volt AC cut down to about 6-8 volts by means of a transformer. The voltage may be varied further by using a 2-watt, 250-ohm rheostat connected into the system. Care must be taken to prevent the wire from burning out from too much current, which is the real purpose of the rheostat. But, even so, burn-outs will be fairly frequent since any wire heated in the open air will burn out sooner or later. This is not a serious deficiency of the apparatus, for you can make up a supply of replacements ahead of time, and it is a simple matter to add a new wire to the binding posts. The wire will not glow brightly—a red heat is the best you can expect without frequent burn-outs —so the testing must be carried out in almost complete darkness, especially for long-focus mirrors.

The Onions light source has several advantages over slits produced by incandescent lamps. It creates rays which spread out evenly over a 360° range, half of which are shielded from the observer's eye by the knife edge. It is simple and easy to use, can be set up at very low cost, and has no optical elements or gadgets to get in the way of the operator. It is a completely new idea, and those who have used it say that its merits outweigh its defects by a large margin. The beginner, however, may find it hard to use and will do well to stick with the ordinary lamp and knife edge apparatus until he gets used to the techniques of testing his mirror.

USING THE FOUCAULT TESTER FOR FOCAL LENGTH

Set the mirror in its holder, first sluicing its surface with water to make it reflective. Place the lamp slit at a distance equal to the radius of curvature of the mirror. Stand behind the lamp and pick up the reflection of the slit on a piece of ground glass at the point of sharpest focus. Adjust the position of the mirror until the image of the slit is within a millimeter of the edge of the knife. While all this is going on, the mirror probably will have dried. Squirt it with water again, a process you will need to repeat several times as the testing goes on. Now place your eye behind, and as close as possible to the knife edge, and move your head until the mirror appears flooded with light. Find the point where moving the knife edge into the cone of light causes the mirror to darken all over evenly. If the knife edge and the slit are not the same distance from the mirror, the average of the two is the radius of curvature. Half this distance is the focal length. Using this method, you can determine the focal length to within a half inch.

Ideally, the Foucault test should be used only with a polished mirror. The fine-ground mirror is not reflective enough to apply the test; and, when you add water to its surface to increase its reflectivity, you lose accuracy because of the uneven drying of the film of water. Nevertheless, within the limits of accuracy you need at this point, the test is very adequate. You have just completed your tester; now is the time to try it out. Check your results with the light test described on page 37.

You can also use the Foucault tester at this point to check the general curve on the mirror. Move the knife edge inside the radius of curvature of the mirror, then slowly cut the cone of light. A shadow will appear on the left side of the mirror and move across its surface. If the leading edge of the shadow is a straight line, the mirror is spherical or close enough to it for our purposes. Check this by cutting the cone of light from outside the radius of curvature. The shadow, this time, will move in the opposite direction, but it should present the same straight leading edge.

It is important to make these tests at this point, which is why we have interrupted our discussion to talk about how to make and use a Foucault tester. If, for example, you find the focal length to be several inches away from what you had intended, you may wish to go back and correct it. This means returning to coarser grades of abrasive to lengthen or shorten the focal length, using the appropriate strokes to accomplish the purpose. Or if, instead of finding a spherical surface on the mirror, you find a deep hole in its center, you must stop to correct this condition before going on to the final stages of fine grinding. The chances are that you will find just what you are looking for, a mirror of the right focal length (or very close to it) with a spherical surface. Old hands at mirror making are likely to sneer at all these tests at this stage of the operation, and it is true that practiced hands can complete the rough and fine grinding stages without ever checking anything but focal length. The beginner, however, is wise to correct any mistakes at this point, for if they are gross errors it will be impossible to correct them later without retracing all steps from the middle of fine grinding on. Furthermore, it is comforting to know that what you have done up to this point is correct and is supported by evidence to prove it. It makes tackling the final stages a job which can be approached with confidence.

CHAPTER 6

Pitch Laps and Polishing

SUCCESS IN POLISHING the fine-ground mirror depends upon two factors: the ability to make and maintain a good pitch lap, and facility in adapting the lap and strokes to maintain a good surface on the mirror during the process. The polishing procedure is a long one, and time and careful work are its essence. We shall describe various short cuts that you may use to speed it up; but, in the long run, the best-polished mirror is one which has been carefully nursed from the opaque surface of fine grinding to the gleaming transparency characteristic of a high and uniform polish.

THE PITCH LAP

In chapter 2 we discussed a few characteristics and types of pitch. In practice, it makes little difference what kind of pitch you use as long as it has the viscosity necessary to make it conform to the surface of the mirror and in addition has the necessary hardness. Dealers supply pure strained pitch as a matter of course, but if you suspect its purity you may want to strain it before using it on a lap. Melt it over an electric plate or Bunsen burner. (Be careful when melting pitch. Do it slowly and cautiously, since hot pitch is inflammable and the material used to dilute it—turpentine—is even more so.) When it is smoking hot, strain it through a few layers of cheesecloth.

Testing for Hardness

The hardness of the pitch lap is of prime importance. If the pitch is too hard, it produces minute scratches, called "sleeks," on the surface of the mirror. If it is too soft, the edge of the mirror will plow into it. The result is what mirror makers call "turned-down edge," a condition which, if present even in mild form on the finished mirror, seriously affects its performance. Soft pitch also makes a lap that is very difficult to maintain. When it warms up from the friction of polishing it has a tendency to "slump" or lose contact with the mirror. Test the pitch for hardness with the Everest thumbnail test (named for A. W. Everest, one of the pioneers of amateur telescope making). Press the thumbnail into the cold pitch with a force of one pound (about the weight of the hand when the elbow is supported on the table) and observe how long it takes to make a dent ¼ inch long. Using this standard, soft pitch will be dented in 5 seconds; very hard pitch in 45 seconds. If the mirror is made of Pyrex it may be worked safely on 40-second pitch, although with plate glass 25-second pitch is the limit.

Tempering the Pitch

Pitch that is too hard may be softened by adding a little turpentine after melting it in an old saucepan. Stir the molten pitch thoroughly with a thin stick—an old egg-beater is even better—add a *few* drops of turpentine and stir some more. Remove a spoonful and drop it in room-temperature water. After 5 minutes, test it with the thumbnail. If necessary, continue adding turpentine until the pitch has softened to the desired limit. Conversely, pitch that is too soft may be hardened by prolonged heating and stirring to drive off some of the volatile material. Try to get the hardness just right for the type of glass your mirror is made of. If you must err, do it on the side of overhardness.

MAKING THE PITCH LAP

A pitch lap is simply a layer of pitch applied to the tool, smoothed to fit the surface of the mirror, and channeled to permit free circulation of air, water, and polishing agents. The channels serve also to furnish space so the pitch can flow to conform to the surface of the mirror. The facets between the channels may be of any shape or size but, for practical reasons in making the lap, they are usually square or round. There are many ways of making laps. We shall discuss several. Choose whichever seems to conform best to your talents. In making any lap, wear old clothes or some sort of protective garment. Pitch and turpentine are ruinous to fabrics, and optical rouge stains even more than that used by milady.

The Sawed Lap

Warm the mirror and the tool by immersing them in a bucket of water heated to a temperature of not over 130° F., while the pitch is heating.

Make a collar of waxed paper and secure it around the tool with an elastic band, overlapping it at the joint. Trim the paper so that it projects $\frac{1}{4}$ inch above the edge of the tool. Prepare a mixture of rouge and water (a tablespoonful of rouge in enough water to make it the consistency of heavy cream), add enough soap flakes to make the mixture feel very slippery to the fingers, and put it aside. Place the tool on a square of aluminum foil (this will make it easier to clean up later) and paint its surface with turpentine. Some workers use hot beeswax for this, but turpentine will make the pitch stick to the tool as well as anything. Now stir the molten pitch to be sure that it is even in consistency. The pitch should be below its boiling point. The stirring helps to force out bubbles that later may mar the surface of the lap. Using a circular motion and starting from the center, pour the hot pitch on the tool until its level reaches the top of the waxed paper band. Leave it to cool.

Now swab the surface of the mirror with the soapy rouge mixture. In about 10 minutes the pitch will have cooled and hardened enough to stand up by itself without flowing over the edge of the tool. Strip the paper collar off, swab the surface of the pitch liberally with the soapy rouge mixture, and place the mirror on it. Work the mirror around, an inch or two at a time, to a variety of positions on the lap to make sure that the surfaces of mirror and pitch conform at all points. This must be done carefully or the edge of the mirror will gouge holes or plow furrows in the still relatively soft pitch. If it becomes necessary—if the mirror starts to stick or bind—add more rouge mixture, but do not lift the mirror off for this. Slide it to one side, add the mixture to the exposed portion of the lap, and slide it back. If the mirror is removed, it is difficult to replace it without trapping air bubbles under it, and these will produce large holes in the surface.

After an hour, or sooner if the pitch seems thoroughly hardened, remove the mirror by sliding it off. Place a paper dinner plate over the lap to keep dust from it. Support the edges of the plate so it does not come in contact with the surface of the pitch, and allow the lap to cool for several hours or, better still, overnight.

Channeling the Lap

When you return next day, trim the edge of the lap with a razor blade or sharp chisel. We have found an X-acto knife an ideal tool for this. But whatever tool is used, wet it well with the rouge mixture to prevent pitch chips from sticking to it. Shave the edge all the way around, using short strokes of the knife. Do this while the lap is still on the aluminum foil, since chips of pitch adhere to anything that they touch.

Find the center point of the lap and mark it with a pencil. Now, using a ruler and pencil, draw lines at right angles to each other in order to produce $1\frac{1}{4}$-inch squares all over the lap. Draw the first two lines in such

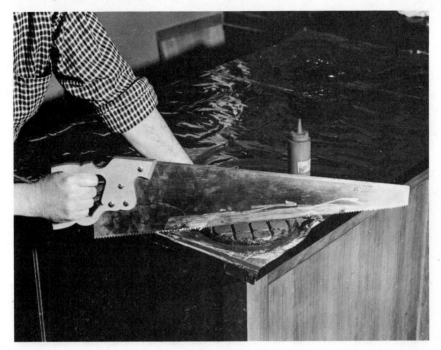

Cutting out the facets.

a way that the *corner* of a square is located at the center of the lap (otherwise the mirror will polish in rings). Soap the teeth and sides of a carpenter's rip saw, then cut along each one of these lines in turn. It is easiest to start with lines of intermediate length until you become accustomed to the way a rip saw acts on pitch. You will have most trouble with the short lines near the edge since the saw will have a tendency to chip out large sections of pitch here (there is less support by other pitch than at the center cuts). Cut clear down to the glass along each line. Turn the lap and saw out the perpendicular lines. Be even more careful here for you are now producing facets with each cut. Irregularities of saw strokes will chip the facets, especially the smaller ones near the edge.

Use the knife to bevel the edges of the facets so that each channel is about ¼ inch wide. Clean up everything when you have finished this job and inspect the lap. All channels should be cut down clear to the glass, and because you deliberately placed the middle facet off center, you will have a row of small facets along one chord of the lap, near the edge. If there are chips here and there along the facet edges, don't worry about them unless they are all concentrated in one area of the lap. In this case, you can build out the facet with a little melted pitch. Similarly, don't be

Trimming the facets. This lap looks rough at this stage, but continued pressing and trimming brought it into shape. It was later used very successfully in polishing and parabolizing an 8-inch mirror. Note the use of aluminum foil to keep the table top clean.

concerned with small bubble holes that appear on facet surfaces unless they too appear only in one area of the lap.

Making Contact Between Lap and Mirror

Apply a thick layer of soapy rouge mixture to both mirror and lap, place the mirror on the lap, and wrap an old wet towel around their joining edges. Place a flat board on the mirror and pile 10 pounds of weights on the board, distributing the weight evenly. Leave for 4 or 5 hours. This is called cold pressing. At the end of this time the surfaces of lap and mirror should be in good contact, with the facets clearly visible when seen through the mirror. If they are not, warm both mirror and lap for 10 minutes in 100° F. water. Paint the lap with rouge and work the mirror over it, using considerable pressure, until you can tell by feel and appearance that there is good contact on all facets. A lap having good contact will produce an even drag as the mirror is pushed across it; there will be no binding or slipping.

Don't be discouraged if the mirror slips and grabs on the lap, even after you have been working on it for 30 minutes or so. Some laps will behave this way for an hour before conforming to the mirror. Patience will conquer any lap. You can tell from the appearance of the facets when you have the lap nearly "licked"—they will all have about the same color as you look at them through the mirror. When you are assured that you have uniform contact with each facet, remove the mirror and inspect the lap to be sure no channels have become plugged with pitch.

Waxing and Screening

Heat some beeswax until it is smoking hot and paint each facet. Use several layers of cheesecloth wrapped around a flat stick, dip it in the hot wax, and paint each facet with a single stroke. Many workers prefer rubbing each facet with a piece of dried paste wax (floor or furniture); but, if this is used, do not try to go over the facets a second time, since these waxes contain a solvent that will make the surface peel away if they are disturbed before the solvent has time to evaporate.

As soon as the solvent has evaporated (or the beeswax has hardened), cut a square of plastic window screen a little larger than the diameter of the lap and lay it over the facets. This can be purchased at most hardware stores and costs about 50 cents a yard. Be sure to buy the woven kind, not the type having the strands fused to each other, for the fibers must slide over one another to allow the screening to adapt itself to the curved surface of the lap. Wet the screening and lap with the rouge mixture, place the mirror on top, and cold press until the pattern of the screening shows in both directions. Strip off the screening. If, because of uneven surface of the screening, there is not perfect contact between mirror and lap when the screening is removed, cold press again for a few minutes and you are ready to use your lap. The window screening serves to add more facets to the lap; and the more facets, the more rapid the polishing action will be. The theory here is that polishing action takes place mostly at the boundaries or edges of the facets. The use of the screen makes a multitude of tiny facets, together having much greater length of boundary or edge than have the usual large facets alone. The idea of using plastic window screening for this purpose comes from E. L. Mason of Portland, Oregon, and is one of the many contributions he has made in the field of amateur telescope making. An alternative to the use of window screening: Use fine-woven nylon netting. It works just as well.

MOLDED PATTERNS FOR PITCH LAPS

Sawing a lap accurately can be a difficult operation. If done properly it produces an excellent lap, which is why it is still used by many amateurs. But for those who have no great proficiency in the use of hand tools it is likely to be messy and time-consuming. In recent years much of the difficulty of this particular phase of telescope making has been removed by the development of rubber molds or patterns. By their use, facets are formed ready-made on the lap. The molds come in two forms, the grid type, which produces square facets, and the so-called button type, which makes round or hexagonal facets. (See appendix for sources of these molds.)

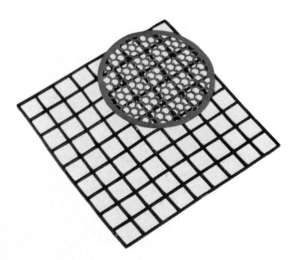

Two types of lap mold. Either one is effective, although the button lap works best with mirrors up to 8 inches in diameter.

The Grid Mold

The preliminary steps for making this type of lap are much the same as for making the sawed lap. Heat both mirror and tool in 130° water. Remove both, then swab the tool with turpentine and the mirror and grid with the soapy rouge mixture. Place the mirror face up on a square of aluminum foil and cover it with the grid, being sure that one corner of a grid square falls in the center of the mirror. Pour the hot pitch on the grid mold. Place the tool, convex side down, on the pitch. Invert the whole arrangement at once and start working the mirror over the grid until all the grid ribs can be seen through the mirror. Slide the mirror off and, as soon as the pitch is hard enough to stand by itself, remove the grid by pulling up on one corner. Trim off the edges of the lap. If, during this treatment, the marginal facets appear to have sunk, the following method is effective in bringing them up to the level of the others.

Scrub the lap thoroughly to remove all traces of soap and rouge. Rub a little turpentine on the marginal facets, then paint on some hot pitch to build these facets up. Use the same kind of cheesecloth swab as described earlier for painting on beeswax. Clean out the channels if any of them shows signs of being clogged.

The remaining steps are the same as for the sawed lap, with sufficient hot or cold pressing to be sure all facets are making good contact.

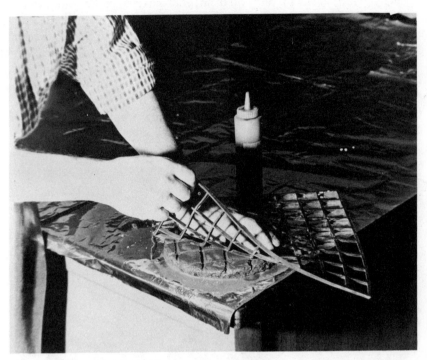

Using a rubber mold in lap making. Don't be discouraged by a knobby appearance at this stage; all laps respond to doctoring and patience.

The Button Mold

This is the famous method written up by Allyn Thompson,[1] and is used widely by many amateurs, including the workers in the optical shop at the Hayden Planetarium in New York City. Thompson made his molds by punching $\frac{3}{8}$-inch holes in a rubber mat $\frac{1}{16}$ inch thick, spacing them about $\frac{1}{8}$ inch apart. The pitch is applied by the method used for the grid mold.

The button type of lap is an excellent one since there is great freedom for the circulation of air, water, and abrasive between the facets. But this type is more applicable to small mirrors than to the larger sizes. Square facets work better on mirrors over 10 inches in diameter. Thompson also graduated the spacing of the facets to make them slightly more concentrated near the margin than in the middle of the lap, thus allowing varied polishing action by using different parts of the lap. This is probably not neces-

[1] See *Making Your Own Telescope*, Sky Publishing Company, Cambridge 38, Massachusetts.

sary, and you can space the facets at equal intervals all over the lap without sacrificing any of the lap's good qualities. There are two ways of punching the holes. In each of them you make a paper pattern which you attach to the rubber with a spot or two of glue. Then punch the holes with an ordinary leather punch which you can obtain from a hardware store, using the sawed end of a 4×4 as a base. Or you can use a drill press to punch the holes, a process which is considerably faster. But punching some three hundred holes in a rubber mat is a tedious procedure, even using a drill press. Buying one of these mats is a good investment for the prospective mirror maker, especially if he plans to make more than one mirror. The 8-inch size costs about $8.00.

A button lap whose effective facet area has been increased through the use of plastic screening. The masking tape is to prevent evaporation during cold pressing. The white line at the bottom of the picture is not a mirror scratch; it is a meaningless defect on the back of the mirror produced during casting.

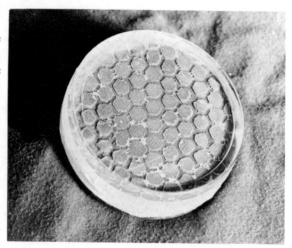

Laps Made from Pre-Cut Pitch

Another way to make a pitch lap is to attach pre-cut squares to the tool in positions already laid out in a penciled pattern on its surface. This is a recommended procedure by some authorities and has the advantage of avoiding a great deal of the messiness and fussy work that go with the other methods, provided the squares are purchased rather than made. (See appendix for source.) They cost $1.25 a dozen for 1-inch squares, ¼ inch thick. Three dozen are needed to cover an 8-inch tool (leaving a few extra for replacements).

Divide the tool into 1-inch squares by drawing lines such that the squares are ¼ inch apart. In this lap, as in all others, the center of the tool must be in the corner of a square. One recommended technique is to heat the tool to a temperature of 150° F. and simply place the squares in their assigned positions. The heat of the tool is usually enough to melt the bottom of the square and make it glue itself fast. A safer method is to heat the tool to a lower temperature (say 110° F.) and cover it with a thin layer of hot beeswax. Warm each square of pitch by holding it near a Bunsen burner or hot plate, then place it in position on the tool, heated side down, and apply a little pressure until it cools. Since the outside facets of the lap are always of irregular shape, squares must be cut to fit. You can do this quite easily by heating a knife hot enough to slice through a square but not so hot that it melts the pitch on each side.

By the time you have finished placing the squares in position the tool will have cooled. Warm it up to 100° F. and hot press for ten or fifteen minutes. The surface of this lap will be more uneven at first than other forms of lap since the facets were not originally formed against the mirror itself. Consequently you may have to hot press several times before the facets make good contact all over the mirror's surface. This amount of hot pressing is likely to deform the facets and you will probably have to do considerable trimming. Follow this by cold pressing for an hour or two.

If you should run out of squares, you can manufacture some more yourself rather than wait a week or so for a fresh supply to arrive. You could, of course, make up the original supply this way. Cut out a dozen strips of wood, each a foot long and ¼ inch square. Place a sheet of paper, waxed or glazed to prevent its sticking to the pitch, on a 14×14-inch square of smooth plywood and tack the strips on this base, 1 inch apart. Leave the nail heads protruding a little for easy removal later. Close in the open ends of the rows by tacking 2 more strips at right angles to the ends of the original ones. This forms a series of channels each an inch wide and a foot long. Heat the pitch, temper it to the desired hardness, and fill each channel. Be careful not to let the pitch overflow. Allow it to cool for several hours. Strip away the wood separators carefully since the pitch strips will be brittle and will crack easily. You now have ten strips, each 1 foot long, 1 inch wide, and ¼ inch thick. Remove the waxed paper from the bottom of the strips and cut each one into 1-inch squares.

Any of the laps described above is acceptable to the mirror maker. The worker may choose the particular one he thinks most adaptable to equipment he has on hand and the amount of time he has available for lap making. This is a task that must not be skimped in any way, for a good mirror demands a good lap. Once the lap is made, take care of it. It is a good idea to keep the mirror on the lap at all times. It protects both the mirror and lap and, most important, guarantees perfect contact between them at the beginning of the next polishing spell. There are risks involved in this procedure, however. The lap must be well covered with a rouge mixture and you must wrap masking tape around the joining edges of mirror and lap, otherwise they will stick together and you may ruin the lap getting them apart. Wrap the disks in wet cloths to prevent evaporation. If the lap is waxed, don't leave them together over long periods of time because the wax contains enough acid to etch the glass slightly.

THE NATURE OF POLISHING

Just what action polishing agents have on a glass surface is still a matter of speculation. Glass, in spite of its hardness and brittleness, is a super-cooled *liquid,* not a solid. It has no definite melting point and no crystalline structure. As temperature increases, glass merely becomes softer. This softening enables the molecules to slide over each other to new positions, just as the molecules of any liquid are free to slide around. The polishing agent heats the surface of the glass, freeing molecules in the areas where most friction takes place and pushing them to new locations. For example, if a pattern is etched on the surface of a mirror—leaving the mirror on a waxed lap will sometimes have this effect—it may be removed by polishing. But if a more stringent polishing action is then applied to the glass, the pattern will reappear, indicating that further shuffling of the molecules has taken place. The pattern, previously filled in by molecules, has been exposed again. This is borne out by experiments which show that iron compounds in minute quantities (rouge is an oxide of iron) can be detected *under* the surface of polished glass.

This conception of polishing is not necessarily in conflict with the idea that particles of abrasive, embedded in the pitch matrix, exert a shaving action on prominences. Possibly the true nature of polishing is a combination of the two actions, reshuffling individual molecules while shaving off large groups of them.

POLISHING OBJECTIVES AND TECHNIQUES

Our first job is to eliminate the pits left by the final stages of fine grinding. When we have done that, we concentrate on polishing out irregu-larities left on the surface of the mirror by defects in the lap or errors in polishing technique. These irregularities consist of hollows, bumps, raised or depressed rings, turned down or turned up edges, lumpy surfaces, etc. They are infinitesimally small according to ordinary standards of measure-ment, projecting only millionths of an inch above the smooth curve we hope for; but, since each affects the performance of the mirror, we must eliminate all of them.

The reader may wonder, at this point, why such tiny defects can have such a serious effect. Our objective, as we have stated before, is to produce a surface that will reflect the parallel rays coming from a distant object to the smallest possible point. But each defect on this surface causes the light rays which strike it to be reflected inside or outside the point. In other words, the rays are displaced laterally from the path they would take upon being reflected from the mirror. Even more serious, there is a very large longitudinal displacement of the point where they meet to form an image. The consequence is that the image becomes blurred or fogged with light coming from the displaced rays. We hope, therefore, to eliminate all defects larger than a certain limiting size, and to produce a curve that is smooth and regular from edge to edge of the mirror. In case this sounds impossibly difficult, let us hasten to add that we have ways of controlling the size of defects so their elimination becomes almost a matter of course.

To start with, we must control the temperature of the polishing room. Both lap and mirror change their characteristics at different temperatures. The mirror expands or contracts slightly, the lap becomes softer or harder, swells or contracts. The surface of the mirror changes with expansion; if we make a test at 68° F. and, a little later, another at 85°, we shall not be testing the same characteristics. Hence all tests must be carried out at the same temperature. For practical reasons we select 68° to 70°, since this is the easiest to maintain. We shall not be too concerned with temperatures during the stages of polishing where we are concentrating only on removing pits; but later on, when we are correcting defects in the surface of the mirror, temperature control is vitally important.

In all corrective operations, almost as much time is spent in waiting for the mirror to cool, and in cold pressing to keep the lap in shape, as in the actual polishing. After 5 minutes of polishing, for example, the surface of the mirror has expanded from the heat generated by the friction of mirror over lap. Not only this, but some parts have expanded more than others if they have been in closer contact with the lap. Consequently we must wait for 10 minutes or so for the mirror to cool before testing it. We may cut this time down by cooling the mirror in water at room temperature, although many mirror makers would frown on this practice. Then, to make sure we have good contact when the mirror is returned to the lap, we must cold press for another 5 minutes.

At the same time, we do our best to control humidity. If the air is very dry, rapid evaporation takes place from exposed facets of the lap. This cools the marginal facets, although at the same time the central facets are becoming warmer because of the friction of the mirror. If the humidity is kept as high as possible, evaporation is kept to a minimum and so is the uneven cooling of the lap. One way to do this is to keep the floor wet, easy to do if you are working in a cellar but not so simple in your wife's kitchen. In this case, flat pans of water placed at strategic spots produce somewhat the same effect.

Polishing to Remove Pits

Make a mixture of cerium oxide or Barnesite with water in the proportions of 1 part of polishing agent to 3 of water. This mixture is kept in the plastic squeeze bottles described earlier. If none is available, use a covered bottle equipped with a large medicine dropper.

Shake the bottle, wait a second or two for the mixture to settle, and make a ring of the polishing agent around the marginal facets of the lap, add a large spot in the center, and gently lower the mirror on the lap. Allow it to rest there for a moment to embed the polishing agent in the surface of the lap, then move the mirror around until every facet is coated. Just how much polishing agent to add at a time is difficult to say, since this depends on many factors—mirror size, facet size, etc. It is better to start with too much in order to get the lap well charged with polishing agent. After this, good practice is to add just enough to keep each facet supplied. Make additions on the watery side rather than a heavy mixture. After an hour's polishing, the facets should appear to be nearly black as seen through the surface of the mirror. If the mirror is making good contact with the lap, the polishing strokes will proceed smoothly and a distinct drag can be felt with each one. If there is any indication of binding or slipping, stop and cold press.

The strokes are the same as those used for fine grinding—⅓ the diameter of the mirror, with the center of the mirror moving from side to side across the center of the lap in the "W" pattern. Most of the polishing will be done with the mirror on top. Later on, to correct a few zonal defects, you may find it necessary to reverse the positions of mirror and lap. Only 20 to 25 strokes (one complete back-and-forth motion) a minute is fast enough for polishing, since rapid action heats the lap excessively. Moderate pressure can be applied to the mirror during all polishing operations. You might read the comments on correcting mirror irregularities (page 87) before commencing actual polishing to aid you in avoiding those errors.

Some mirror makers advise attaching a handle to the back of the mirror during polishing to prevent the heat of the hands from being transferred to the glass. But attaching a handle introduces other errors just as bad, such as "rocking" the mirror during a stroke. This causes the leading edge of the mirror to dig into the lap if it happens in the middle of a stroke, or makes the mirror tip over the edge of the lap at the end of a stroke. Both produce effects far worse than those of the extra heat, and so a handle is not recommended. But be careful not to let the finger tips curl over the edge of the mirror because their warmth, applied so close to the surface of the mirror at its edge, will produce small local bumps on the glass in this area. The bumps are polished off and later, when the mirror cools and contracts to its normal dimensions, become hollows instead of hills.

At this beginning stage, however, you are interested chiefly in removing pits left from the final stages of fine grinding. So polish away, stopping only to add a fresh charge of polishing agent if necessary. You need not remove the mirror for this. Slide the mirror back far enough to expose a few marginal facets, add the polishing agent to them and then do the same on the other side.

After a half-hour spell of polishing, wash and dry the mirror and inspect its surface. The mirror should appear semipolished over all of its surface, that is, a "shine" now begins to appear on the fine-ground surface. This is the ideal condition at this point, for it means that continued polishing will soon get the job done. But if the polish shows up unevenly, there is still nothing to worry about, at least for a while yet. For example, the center of the mirror may appear semipolished, while the edges show only a slight difference from their previous fine-ground appearance. This is because the edge facets have a tendency to "sink" faster than those at the center, and also because the center of the mirror is always in contact with the lap and its edges are in contact for only part of the stroke. If the lap is hard enough, the polish will spread to the edge during subsequent polishing spells and the mirror will then polish all over.

Cold press for 10 minutes after returning the mirror to the lap. This is necessary because both mirror and lap become heated during the polishing period but, as soon as they are separated, cool at different rates. They are then no longer in perfect contact. Even if your endurance is such that you can polish continuously for hours, it is not advisable to do so. During polishing, some facets swell more than others. A facet which has swollen so that its surface is even slightly above its neighbors exerts more polishing action than they do. This will affect the part of the mirror immediately above the facet, and a zone is produced. Consequently, you waste your time if you polish more than 20 or 30 minutes because of the repair work you will have to do later.

When you leave the lap overnight, cover it with the mirror as suggested earlier to insure good contact between mirror and lap when you start to polish next day. Don't forget to put masking tape around the edges of the disks and wrap them in wet cloths to prevent evaporation. If the polishing agent between the disks dries out overnight, you may have to ruin the lap to get them apart. One way to cut down evaporation is to add a little glycerine to the final wet.

After two hours polishing, the mirror may appear to be perfectly polished as seen with the naked eye. But if you examine the surface with a magnifier (see page 48), you will be able to spot many tiny pits in the vicinity of the reflection of the light. These will be more numerous near the edge than at the center of the mirror. You must continue polishing until the last one disappears.

The lap must be constantly checked. If the channels begin to close up, clean them out with a knife. If the pitch spreads out around the lap margins, trim it away. Watch the pattern of the screening material on the facet surfaces. When it becomes indistinct, warm the lap in hot water, then hot press with the rouged screening between mirror and lap until the pattern reappears. A faster method is to warm the lap surface with a heat lamp for about 2 minutes, moving it constantly about 12 inches from the lap. Then press with the screening. These measures will probably be unnecessary because a well-made lap of the proper hardness will last through the entire polishing and figuring processes.

It sometimes happens that, even after several hours of polishing, either the center or the edge of the mirror refuses to take on a polish. If this is the case with your mirror, you must return to fine grinding, starting with No. 400 abrasive. Each of these defects is a hangover from too long a stroke in the fine-grinding process. If the center refuses to polish, the mirror is hyperbolic (see page 96); if the edge is recalcitrant, it is because it is badly turned down. Neither can be corrected by polishing if it doesn't respond in the early polishing stages. Furthermore, prolonged polishing may itself damage the mirror, producing a surface which will be very difficult to correct later on. The only remedy is to grind the defects out with abrasive, using the ⅓ "W" stroke. The chances of finding either one of these defects is very slight, fortunately, if a thorough job was done in the original fine grinding.

When the last pits near the mirror's edge have finally disappeared, the preliminary steps of polishing are complete. The surface of the mirror is probably very irregular from the point of view of its optical qualities. But seen with the naked eye, it is a thing of beauty.

QUICK-POLISHING METHODS

The methods and techniques we have been talking about in the previous section will produce a well-polished mirror in 5 to 8 hours of actual polishing time.

There are two other methods which will produce the same result in considerably less time; but, like all short cuts, they are attended by risk. The fast-polishing action can produce some undesired results somewhere along the line, requiring a consequent retracing of steps; but so can the most orthodox methods when used improperly. For those who find long periods of circling a barrel wearisome, these short cuts may be attractive enough to warrant the risks involved. Lest you be frightened away from short cut polishing by this warning, let us hasten to add that the risk is small indeed if reasonable precautions are taken. Each method comes from E. L. Mason. We have tried them at Millbrook with complete success and have no hesitancy in recommending them. As a matter of fact, it is very likely that Mason's plastic lap method will become "the" method for polishing mirrors. It is easy to set up, rapid in action, and produces a polish comparable to that obtained on pitch.

Polishing with Abrasives

Make a hard (40-second) lap, facet it, and cold press to good contact. Don't wax the lap; the pitch facets work better by themselves.

Dust on the finest abrasive you have (not over 4 to 6 microns; Bausch and Lomb's No. 2600 or 4-micron garnet powder) and spread it with the finger tips until each facet has a smooth, even layer of abrasive. Spray with water from the atomizer bottle. Place the mirror on the lap and press down strongly to embed the abrasive in the pitch. Spray again and start to polish, using the same stroke as for fine grinding. Add more abrasive if necessary, and polish for 45 minutes to an hour, stopping to cold press if the lap deforms because of the heat produced. It is essential in this method that good contact be maintained at all times, since a malfunctioning facet covered with abrasive can produce weird effects on the mirror. At the end of an hour's time, probably less, the mirror will be semipolished all over and can be tested with the Foucault apparatus to see what has happened to the surface.

If the curve is all right, you can make another lap to complete the polishing process. In case making another lap seems arduous, remember that you have saved from 2 to 4 hours of polishing time.

Now to add a short cut to a short cut. You can proceed with the *same* lap if you want to. Scrub it under the faucet, then paint it all over with hot beeswax to cover up the abrasive. If you are cautious to the extent where you don't trust wax to cover abrasive successfully, paint the facets with a thin layer of pitch first. If you wish, you can use any microcrystalline wax as a substitute for the beeswax (see page 69). After waxing, cold press for sufficient time for the well-rouged mirror to make complete contact. Break up the facet surfaces with plastic window-screen material, cold press again, and you are ready to complete polishing. You can finish the whole polishing job, from start to finish, in about 6 hours for an 8-inch mirror.

THE PLASTIC LAP

Surfaces other than pitch have been employed for many years in polishing mirrors. Paper, silk, beeswax (in the form of honeycomb foundation), and cloth have been widely used, but all have the same defect—they produce a relatively unyielding surface which is hard to manage, and also have a tendency to produce a "lemon-peel" surface on the mirror. Mason discovered, however, that, if a pitch lap is covered with a fine-grained plastic material (the kind that has small projections embossed on its surface, see appendix for sources), the lap still retains all the qualities of its pitch foundation and in addition polishes a mirror at least twice as rapidly as the pitch alone.

Make the usual hard pitch lap and facet it. There is no need to wax the lap; the plastic works better on a pure pitch base. Warm the tool in hot water until the pitch is soft, then lay on a sheet of the plastic

Plastic material useful for fast polishing. Note the projections seen through the eyepiece. Each one increases the polishing action.

Plastic material applied to a button lap.

material. Trim the plastic to within ¼ inch of the edge of the lap and cold press overnight. The plastic will take the shape of the lap. The mirror does not need to be soaped or rouged for this since it will not stick to the plastic as it does to pitch—this is another advantage of this type of lap.

The plastic surface polishes the mirror quickly with rouge or cerium oxide, but its action is even more rapid with Barnesite. Cover the surface with a Barnesite solution to which you have added a little detergent, and start polishing. Using the standard ⅓ "W" stroke, and stopping to cold press every 15 minutes or so, you can bring the mirror to a full polish in 3 hours by this method.

The mirror will emerge from this process in a slightly lumpy condition if you try to work too fast. The stroke, too, must be adjusted to the way the polish comes up. Slow polishing at the edges of the mirror and the appearance of a central depression, for example, indicate too long a stroke.

Now for the precautions against the "risks" mentioned earlier, which turn out to be not so serious if you guard against them.

1. *Avoid scratching.* You must be even more careful using plastic than you are with pitch. A particle of abrasive or dust dropped on plastic cannot sink into the surface as it can in pitch and therefore does more damage. Be cautious about placing the mirror on the lap, this is when scratching is most likely to occur.

2. *Keep the lap wet.* The polishing particles have less room to move on the plastic. More heat is produced than on pitch because the drag is so heavy and the lap dries out rapidly. When this happens, a clump of dried particles can produce a disastrous scratch.

3. *Use slower stroke speeds.* This lessens the heating effect from the drag and produces a smoother surface. Because of the intimate contact of lap and mirror, slow speeds (half of those on plain pitch) are much more effective than on pitch. They also reduce the tendency for the lap to dry out.

4. *Facet the lap as usual.* Don't assume, because the facets are covered by the plastic, that making a good lap isn't necessary. It is just as important for the pitch to have room to flow with this lap as with any other.

5. *Watch the surface.* If a tear or break occurs in the plastic, replace it at once. Particles pile up rapidly around a break, the liquid drains off into the channel below, and the result is "sleeks" (tiny hairlike scratches) or worse.

6. *Use a plastic whose surface projections are of a size proportional to the size of the mirror.* Fine-grained plastic is good with 6 and 8-inch mirrors. Coarser ones work better with larger mirrors be-

cause the drag per square inch is less. A large mirror on a fine-grained plastic is almost immovable.

7. *Don't expect the plastic to do the whole job.* For the final stages of polishing, strip off the plastic. You will find a clean pitch surface underneath, marked with the pattern of the plastic. If the pattern is clear and distinct, cold press with the mirror until the contact is perfect. But if the pattern is spotty, remove it by warming the lap either in hot water or by the heat lamp mentioned earlier. Then press a window-screen pattern into the pitch. Replace the Barnesite that you have been using with fine washed rouge. Be sure the lap is making good contact everywhere on the mirror and polish for a half hour, stopping to cold press if necessary. At the end of this time you should have a lustrous finish on the mirror.

Each of the above methods—prepolishing with abrasives, doing the whole job with a pitch lap alone, or polishing with a plastic lap—will give excellent results. Each can be mastered by the beginner. The plastic lap method is the cleanest, fastest, and most efficient for taking a mirror from the fine-ground stage through the polishing operation and usually is the preferred method by workers who have tried all three. On the other hand, the old tried-and-true method, while time-consuming and hard on the patience, still has merit for those making their first mirrors. The very slowness of the process is its great virtue because you can observe constantly what is happening to the mirror. No defect will reach a stage where steps will have to be retraced when you work slowly, which means that most of your corrections will be preventive rather than remedial.

CHAPTER 7

More About Polishing

WHEN YOU HAVE FINISHED polishing the mirror, the gleaming surface and the smooth curve will make a non–mirror maker wonder what else could possibly be done to improve it. Very little, you hope; but, in order to discover what defects may be concealed by its shining face, you turn to the Foucault test apparatus.

USING THE FOUCAULT TEST TO LOCATE MIRROR DEFECTS

You no longer need to wet the surface of the mirror to make it reflective; its own polish takes care of that very effectively. But when you test the mirror now at the center of curvature, instead of darkening evenly all over, it appears covered with shadows of various shapes. The interpretation of these shadows is the key to eliminating the defects that cause them.

As we have said previously, a perfectly spherical mirror will reflect rays from the lamp in the Foucault test apparatus back to a point. If the eye is placed at this point, the mirror appears flooded with light. If the knife edge cuts the cone of light here, the mirror darkens all over. But if the surface of the mirror is not perfectly spherical, what then? Some rays will be reflected back to the point but others, striking irregularities in the mirror, will be reflected in whatever direction the slope of the irregularity dictates. The knife edge will cut off some of these rays, but the others will by-pass it and still strike the eye. Thus some areas of the mirror will appear to be very dark, others will be illuminated, and still others—where part of the rays from a particular irregularity are cut off but some still get by the knife-edge—will be gray. Now because a very small defect on the mirror will produce a very large deviation in the direction of the reflected ray, the defects appear to be greatly magnified and even the smallest ones become extremely prominent. This magnification of defects is an astonishing phenomenon. It depends on the focal length of the mirror—the longer the focal length the greater the magnification—and is in the order of 100,000 times. But it is in *depth* only, and the *width* of a defect is seen exactly the same size as on the mirror itself. A bump on the glass only 2 *millionths* of an inch high will appear, under the knife-edge test, as though it were 2 *tenths* of an inch. But if the bump is an inch *wide* on the mirror, it will also appear an inch wide under the knife edge. A striking example of this phenomenon is to place your thumb on the mirror for a minute or two and then measure the bump raised by the thermal expansion as seen under the knife edge.

When the knife edge is placed on the observer's left, cutting into a cone of light reflected from a lamp on his right, the mirror seems to be illuminated by a light source whose rays are striking the surface at a very small angle from the right of the mirror. Protuberances on the mirror will therefore seem to cast a shadow which falls on the *left* of the protuberance and its *right* side will be brightly illuminated. On the other hand, depressions will have shadows on the *right* side, but the *left*-hand limit of the depression will be illuminated. But here the resemblance to real shadows ends, for the shadows never extend beyond the defect that causes them; only to the extremity of the defect itself. Thus we may clearly recognize holes, bumps, raised zones, depressed zones, or other distortions on the surface of the mirror, as indicated in the diagrams. However, we must not get the idea that any shadow indicates a distortion. Some of them—and these are characterized by definite patterns which we shall learn to recognize—are indications of regular curves (other than spheres). One of them, the parabola, is a curve whose shadows we shall encourage rather than try to remove. (See the shadow patterns in chapter 8 and compare them with those illustrated here.)

Recognizing the nature of a defect on the mirror gives us the key to how to treat it. If we see a central bump on the mirror, common sense tells us that if we change the stroke or the lap so that this area gets more polishing than other parts of the mirror, the bump should be ground away. And this is just what we shall do in all cases, always working toward producing what appears to be a *flat* surface on the mirror under the knife edge, for this means that we have really produced a *spherical* one. In deciding what defects to correct we adjust the knife edge to the position where most of the mirror's surface will *appear* flat when the cone of light is cut. It would be a useless waste of time and energy to attempt to adjust the whole surface of the mirror to any predetermined radius of curvature, and to bring all irregularities down to this particular knife-edge setting. This is most important in all phases of testing and correcting the mirror, and it means that during the final operations on its surface we may change the radius of curvature and therefore its focal length a dozen times. These will always be very small changes, involving only a fraction of an inch in the focal length. If we attack each problem in this way, we shall always be removing only a minimum amount of glass in any one operation. This relieves us of a vast amount of work in the long run.

OTHER TESTS FOR MIRROR DEFECTS

The Foucault test is most effective for irregularities in the intermediate zones of the mirror, and is least useful for those in the center or near the rim. So we shall take a moment to discuss one or two others which will help us in these areas.

The Eyepiece Test

This is a classic and infallible test for one of the most common mirror defects, turned-down edge. Substitute an eyepiece of about one-inch focal length for the knife edge. You must also change the slit in the lamp arrangement into a pinhole. This is easily done by slipping a piece of aluminum foil behind the razor blades which form the slit, then pricking its center with a small needle. Only a very tiny hole need be made, and you may use up a considerable amount of foil before you get one that is satisfactory. It must be very small, and as nearly round as you can make it.

Pick up the reflection of the pinhole on a piece of ground glass or white paper at the center of curvature, then move the eyepiece into this position. At the center of curvature, the eyepiece presents a sharp image of the pinhole. Now slide the eyepiece stand inside the center of curvature. If no turned-down edge is present, the image of the pinhole will be a circular disk with a sharp, clean edge. But if the edge is fuzzy or hairy, the edge of the mirror is turned down. We talk about remedial treatment for this defect below.

The Ronchi Test

Another test for turned edge is the Ronchi test, and this arrangement may be used to test other parts of the mirror as well. Here we substitute a grating for the knife edge. A grating consists of a number of fine lines, about 100 to the inch, engraved on glass or photographed on a strip of film. They may be obtained commercially for about a dollar (see appendix for suppliers) or can be made at home by winding fine wire on

Ronchigram of a spherical mirror. Note the slightly turned edge. (Photo by Samuel Borrello, Syracuse, New York)

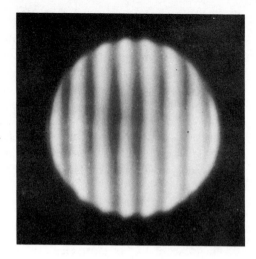

Ronchigram of a parabolic mirror. (Photo by Samuel Borrello)

This is the same mirror as in the previous photograph. The Ronchi grating has been moved closer to the focus. (Photo by Samuel Borrello)

a frame. This is a tricky operation but it can be done by anyone with good eyesight and patience.

How to Make a Ronchi Screen

From ⅛-inch aluminum sheet, saw out two squares 1½ inches on a side. Drill a ½-inch window in the center of each square. Bind the two squares together with Scotch Tape. Using the finest wire obtainable (old radio transformers are excellent sources for this), wind a *double* strand continuously around both sides of the square to cover the windows, laying the double strands side by side. This must be done so that the wires are tight against each other. If done properly, when the windows are covered they will be almost opaque to a bright light held in back of them. Anchor the beginning and ending of the wires with small screws secured to only one of the plates. Now unwind one wire of the double strand carefully so that the position of the other wire is not disturbed. Cement the remaining wires (use Duco cement or airplane glue) to the sides of the squares, top and bottom, back and front. File the wires through at top and bottom, separate the squares, and you have two Ronchi screens in which the distance between wires is exactly the same as the diameter of the wires themselves. You will use one, and keep the other for a spare.

Using the Ronchi Test

Replace the knife edge with one of the screens, narrow the slit between the razor blades in the lamp source until it is about the width of two sheets of typewriter paper, and make sure that the lines on the Ronchi grating are parallel to the sides of the slit.

Pick up the reflection of the slit, and slide the Ronchi screen into the cone of light about ½ inch in front of the radius of curvature. You will see the mirror covered with a series of fine straight lines. These lines should run off the edge of the mirror without any bending. But if they bend inward (toward the vertical axis of the mirror) the edge is turned down. If they bend outward, the edge is turned up. Slide the screen back until it is just inside the center of curvature. The Ronchi screen now behaves like a knife edge in that any irregularities on the mirror show up as distortions of the lines.

CORRECTING MIRROR IRREGULARITIES

Turned-Down Edge

If you have been using ⅓ strokes during the preliminary polishing period, the chances of finding a turned-down edge are not very great. But there may be some instances where you have inherited this defect from fine grinding. If so, concentrate for a ten-minute period of working on ⅓-stroke, center-over-center polishing. Do this vigorously, applying pressure but not hurrying the strokes. If, when you check the edge after this, the turned edge seems to be diminishing, continue with 5-minute polishing periods (cold press between each one) until the defect disappears. If not, shorten the stroke to ¼ diameter and go on with 5-minute periods until you have licked the problem. Unless the turned edge is of gross proportions—in which case you will have to return to fine grinding to remedy it—short strokes and a hard lap will always bring a turned edge back to a good one.

The Foucault test may be used for turned edge if you have a good enough eye to detect it. If the mirror shows a bright, very thin line all the way around its rim when tested at the center of curvature, the mirror has no turned edge. If the mirror is perfect all the way to the rim, the ring on the right side of the mirror will be just a little brighter than that on the left. But if they are equally bright or if the left half is brighter than the right, the edge of the mirror is turned *up*. If there is no ring at all on the left and that part of the mirror is in shadow, the edge is turned down.

Turned-Up Edge

This is seldom very serious and can usually be eliminated by lengthening the stroke for very short periods—about 3 minutes. If a few spells of this action is not effective, don't continue it because the long stroke will also dig a hole in the center of the mirror. Instead, reverse the mirror and the lap and polish with the center of the lap directly over the edge of the mirror. Go easy with this or you will produce a turned-down edge.

Dog Biscuit

A bumpy, uneven surface spread over the whole mirror is usually called a "dog biscuit." It is invariably the result of too-rapid polishing strokes or too-long periods of polishing. The lap becomes distorted as it is heated, the facets expand unevenly, and individual facets heat localized areas of the mirror. These areas expand in their turn and, as they swell out, receive more polishing action than the remainder of the surface. When the mirror is removed from the lap, the heated parts contract and a bumpy surface is the result. The obvious cure is to slow the stroke speed and to make sure that lap and mirror make good contact from the beginning.

A more localized type than dog biscuit is a "lemon peel" surface, usually caused by overheating of the lap but more often the result of specialized laps, such as the plastic lap described earlier. Slowing the stroke, using more variation in strokes, or decreasing the pressure are all remedies for these conditions. The lemon-peel surface, sometimes referred to as microripple, is a minor defect and, at worst, simply causes a little scattering of light.

CENTRAL HILL WITH HOLE IN ITS CENTER
USE BLENDED OVERHANG STROKE TO REMOVE.

"DOG-BISCUIT" SURFACE—THE RESULT OF
WORKING TOO FAST. COLD-PRESS AND TRY
AGAIN.

RAISED ZONE: USUALLY CAUSED BY HIGH
FACETS IN ONE AREA. COLD-PRESS, FOLLOWED
BY 1/3 "W" STROKES.

DEPRESSED ZONE: PROLONGED COLD-PRESSING
FOLLOWED BY 1/3 "W" STROKES, OR USE A
SUB-DIAMETER LAP.

Common mirror defects and their treatment.

Central Bumps or Hills

Find the knife edge setting at which the central bump projects least above the flat appearance of the remainder of the mirror. This will give an estimate of the amount of glass to be removed (see page 84). After operating on the bump, return to this knife-edge setting to check on your progress. The easiest way to remove a bump or hill in the center of the mirror is to lengthen the stroke, but this involves the danger of turning the edge. A compromise between the long strokes and the normal ⅓ stroke is what the dean of amateur telescope makers, A. W. Everest, calls the "blended overhang." In this stroke the mirror overhangs the lap and is worked back and forth with its center near the edge of the lap, as shown in the diagram. It must be used cautiously, for vigorous action will soon turn the bump into a hole, or a bump with a hole in the center. When this stroke is used, polish for very short periods—not more than 2 or 3 minutes. If a hole develops in the bump, extend the stroke a little to widen the area receiving most of the polishing action.

Another way to deal with a central bump is to raise the central facets a minute amount to provide more polishing action. Cut out a circle of plastic screening the size of the bump, place it on the central facets, and cold press for 15 minutes. The pitch will be forced up through the spaces in the screening, thus presenting a slightly higher surface in the center of the lap. Use this treatment cautiously—its action is drastic. Be sure to cold press afterward.

Central Depression

If the central depression is the result of measures taken to remove a bump or hill from the same area, working on a normal lap with ⅓ center-over-center strokes will usually bring the area back to normal. If not, the central facets may be slightly depressed by cold pressing with a circle of waxed paper placed over the offending spot. A very slight depression of the

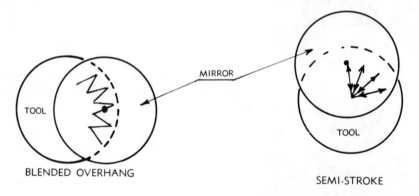

Two strokes useful in polishing.

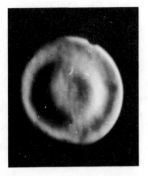  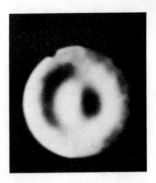

Three stages in the treatment of a 3-inch mirror, starting with an oblate spheroid whose central hill has a hole in its center. *Below,* mirror has "flattened" to a nearly spherical surface. (Focograms by Samuel Borrello)

facets is all that is necessary, just enough to reduce their contact with the mirror. You can tell immediately whether you have reduced the contact by looking through the mirror; facets not making contact have a murky or cloudy appearance. Polish for a few minutes using ⅓ stroke. If the depression does not seem to respond to this treatment, or if other equally bad effects are produced as a result, try a sub-diameter lap.

Sub-Diameter Lap

A sub-diameter lap is always used with the mirror face up. The action is, of course, more localized than that of the full-sized lap, which makes it very valuable in treating defects. Many workers prefer it to deforming the lap by pushing down or raising individual facets. The deformation usually lasts much longer than you want it to and results in a mirror defect opposite in nature to the one you have been treating. Deforming the lap always involves an expenditure of time way out of proportion to the amount of work accomplished and is a specific only for a particular defect.

The sub-diameter lap, on the other hand, can be varied in its action to treat almost any defect.

Making a Sub-Diameter Lap

This can be made by using a plywood or hardwood base, well shellacked to prevent absorption of moisture. It should be about ⅔ the diameter of the mirror; in our case, 6 inches wide. Button facets are best for this lap, and it is made exactly the same way as an ordinary button lap except that the wooden base should be warmed with a heating lamp or hot pad instead of being immersed in water. If you do not have a button-lap mold available, cover the wood base with pitch and press it against the face of the well-rouged-and-soaped mirror. After the pitch has set, cut 1-inch facets (¾ inch if the lap is smaller than 6 inches) with the carpenter's saw.

Using the Sub-Diameter Lap

You can treat either raised or depressed zones with this lap. If the zone is raised, adjust the lap strokes to work directly on the zone. Short, straight strokes are best for this. In depressed zones, use strokes whose action is concentrated on the high areas around the zone. Small elliptical strokes work best here. Remember that a sub-diameter lap works faster than one of full size, so make your polishing time short and test often. When the rim of the depression becomes more shallow, resume work with the full lap, using short center-over-center strokes.

Rings

If a ring, either raised or depressed, appears at the beginning of polishing, it is an indication that the middle of the center facet has been placed too near the center of the lap instead of having one of its corners in this position. The only thing to do in this case is to remake the lap. But if a ring appears after polishing has been going on for some time, it indicates a raised or depressed facet, or group of facets, on the lap. Prolonged cold pressing (twice the usual time) will usually take care of this situation. If it doesn't, try elliptical or figure-eight strokes for a few minutes. Be careful, because these strokes have a tendency to turn the edge. Raised rings are gotten rid of easily, but depressed rings are more stubborn. If a depressed ring persists in spite of the action recommended above, use a sub-diameter lap. Apply a slight pressure to the edge of the lap and adjust the stroke so that the action takes place on the high areas on either side of the ring. Be very cautious here; this can lead to worse trouble unless you make constant checks on what is happening. As soon as the ring shows signs of responding, return to the full-sized lap after cold pressing for contact.

It is impossible to prescribe remedial measures for all the irregularities which may appear on a mirror, and it must be emphasized that the measures mentioned above are only suggestions and not infallible remedies.

Laps are unpredictable tools; just a slight difference in hardness, the way they are faceted, the strokes used on them, the polishing agent used, or any one of a half dozen different factors may change their behavior completely. Some authorities, in spite of this, set up a prescribed treatment and insist that the mirror will come out all right if the treatment is followed slavishly. We have found that common sense is a much better guide; if you think of the lap condition and stroke action in terms of what each is doing to the surface of the mirror, your troubles will be minimized.

Remember that any surface, provided that it is resilient and is not hard enough to scratch the mirror, will serve to reduce irregularities. The thumb, for example, covered with an adequate supply of polishing agent, will reduce a small bump or smooth the edge of a depression. But the rouged thumb makes a small polisher of very drastic action and must be used very cautiously. Otherwise you will find that the bump you treated with it has become a large depression. The lap itself, if worked on top of the mirror, will act as a local polisher if a slight variation of pressure is used to make one area press down harder than the rest. Most of the troubles we have been describing will not happen to your mirror; but, if you do run into a few difficulties, remember that whatever happens to a mirror because of polishing can be remedied by more and different polishing, provided it is tempered by thought.

FINAL POLISHING WITH WASHED ROUGE

Let us assume now that you have reduced all zones to a minimum and all that you can see on the mirror are traces of minor imperfections. If you now change to a long stroke with washed rouge, the remaining irregularities will disappear as if by magic, provided your lap is well pressed; and you can enter the final stage, parabolizing, with a mirror which will need little further attention.

Washed rouge is easily made and well worth the trouble. Add a couple of ounces of rouge to a quart jar full of water containing a few drops of liquid detergent. Shake thoroughly and allow to settle for 15 minutes. Siphon off the liquid from the middle part of the jar into another container and allow this to settle for 24 hours. Siphon off all the liquid you can from this second jar, being careful not to disturb the settled rouge. What remains in the bottom is washed rouge. It is very fine, is slow acting (compared to cerium oxide or Barnesite), and will give you smooth, scratchless polishing.

Using a solution of the washed rouge on the lap, change the stroke for the final polishing after a prolonged session of cold pressing. Instead of the $\frac{1}{3}$ "W" stroke, center the mirror, then slide it up to 12 o'clock, back to center, then to 1 o'clock, and so on all around the lap. This is what its originator, Everest, called the semistroke. It has the advantage of maintaining uniform pressure on all parts of mirror and lap and produces a zone-free action. The final result of all this is that you have a mirror with a uniform surface. It may not be a sphere, but it will be what mirror makers call a figure of revolution (compare yours with some of the drawings in the next chapter), and you can go on to the final step with confidence that you have the situation well in hand.

CHAPTER 8

The Mirror: Testing for the Paraboloid

FIGURES OF REVOLUTION

THE POLISHED MIRROR is free of bumps, holes, and other defects and has a smooth, velvety appearance all over its surface. But there still may be broad areas of light and shadow whose shape and depth seem to depend on where the cone of light is cut with the knife edge. If such shadows are present, they probably represent not local defects in the mirror but one type of the several possible curves which polishing action produces.

These curves are known as figures of revolution because they represent the curves generated by various geometric figures revolving about their axes. Any text in analytical geometry describes these curves in detail. The geometric figures are called *conic sections* because each is a figure produced by cutting a cone in a certain direction.

The *ellipse* is produced by cutting a cone so that we slice through only the curved part of the cone (above the base) at an oblique angle to the perpendicular running from the point of the cone to its base. But if the slice is at right angles to the axis (or parallel to the base), the curve produced is a *circle*. Now if the cut is deepened, so that we slice through the flat bottom of the cone, and also cut in a line parallel to the opposite side, the result is a *parabola*. Finally, if the slice is *not* parallel to the opposite side but passes through the base, the figure is a *hyperbola*. Each of these two-dimensional curves produces a three-dimensional figure, or hollow, when rotated, and these hollows differ slightly in their characteristics when seen under the Foucault test. The three-dimensional figures, to differentiate them from those of two dimensions, are named by adding the suffix "oid." Thus a parabola, when rotated, will produce a paraboloid;

Conic sections. *Top,* the circle; *center,* the ellipse; *right,* the parabola; *left,* the hyperbola.

a circle, a spheroid; and so on. Before we leave these figures we must add one more comment. With a given cone, the only figures where the *shape* remains constant, although the *size* may vary, are the parabola and the circle. On the other hand, there are fat ellipses, thin ellipses, large ones and small ones. With circles and parabolas cut from the *same* cone, we can have variations only in size, not in shape.

If an ellipse is rotated about its major (long) axis, it produces an ellipsoid, or prolate spheroid. The radius of curvature of this figure is *less at its center than at its edge.* Put in another way, *the curve at the center is deeper than it is near the edge.* The circle, when rotated, produces a sphere whose radius of curvature is constant from edge to edge. The oblate spheroid, produced by rotating an ellipse about its minor (short) axis, reverses the trend. The radius of curvature here is greater at the center than at the edge; the curve is shallowest at the center. The paraboloid is even deeper at the center and the hyperboloid is deepest of all.

The only one of these figures which appears *flat* under the knife edge is the spheroid. Unfortunately for testing purposes, all the others produce the same doughnut-shaped figure (toroid) and so it is difficult to distinguish one from the other by *general appearance* alone. But *they all differ in the position of the shadows* as these shadows appear and disappear for the various knife-edge positions, and this is the key to the testing system we shall use. In essence, we move the knife edge to various positions and measure the intensity and position of the shadows on the mirror for each position of the knife edge.

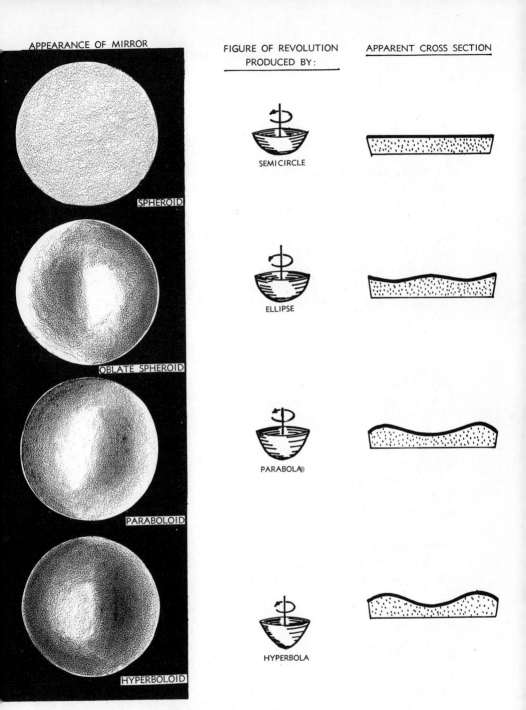

APPEARANCE OF MIRROR	FIGURE OF REVOLUTION PRODUCED BY:	APPARENT CROSS SECTION
SPHEROID	SEMICIRCLE	
OBLATE SPHEROID	ELLIPSE	
PARABOLOID	PARABOLA	
HYPERBOLOID	HYPERBOLA	

The figures of revolution.

Consider the shadows on this page as being caused by light coming in from the right. Then, if you turn the page upside down, the oblate spheroid will look like a paraboloid, and vice versa.

TEST FOR THE PARABOLOID

In the following chapter we discuss methods for deepening a spherical surface into one whose curve is a paraboloid. Because you must test the mirror constantly during this process, we stop for a moment now to discuss the principle of the test used.

Fundamentally, all we do is find the radius of curvature of several concentric circles, or zones on the mirror. The reason for this is that the curve we deal with, the paraboloid, is deeper in the center and flattens out as it approaches the mirror's edge. By testing the radius of curvature at various points along the curve, we can find out exactly *how much* the curve flattens. We then compare our figures with how much it *should* flatten if it is a true paraboloid.

The subsequent sections of this chapter deal with the actual measurements and the theory and reasons for each one.

MATHEMATICS OF MIRROR MAKING

So far we have intentionally avoided mathematical discussions, but since we now must make actual measurements we must also consider the mathematical means of evaluating them.

The curves we deal with are extremely shallow ones, differing so little from one another that there is no possible mechanical means of telling one from another. The parabola almost, but not quite, fits a circle of the same radius of curvature. If these two curves, each having the same radius of curvature at the center, are placed together, the arms of the parabola will gradually move outside the circle. The actual difference in the two curves, for any point on their radii, can be found from the formula $r^4/8R^3$, where r is the radius of the surface at any point, and R is the radius of curvature of the central zone.[1]

Suppose we measure these differences at several points from center to edge, using an 8-inch mirror whose radius of curvature is 112; in other words an f/7 mirror. We choose the points in the table here to make the measurements, for reasons which appear later.

The figures indicate that the difference at the point of widest separation of the two curves (the edge) is about 23 millionths of an inch. They also indicate a fact mentioned earlier; the paraboloid has its sharpest curve at its center, a curve which grows shallower as it approaches the edge of the mirror. The right-hand column in the table shows how much

[1] The formula $r^4/8R^3$ represents only an approximation of the true difference between the curves. More accurate, and possibly easier to use, is the expression $R - \sqrt{R^2 - r^2} - r^2/2R$, where $R - \sqrt{R^2 - r^2}$ is the formula for the sagitta of the sphere and $r^2/2R$ represents that of the paraboloid. This more exact formula should be used for short-focus mirrors; for the 8-inch f/7 mirror, $r^4/8R^3$ is quite satisfactory.

Distance from Center of Mirror		Difference in Millionths of an Inch Between Circle and Parabola
(in inches)	(as per cent of total radius)	
0 (center)	0	0
1.26	31.6	0.224
2.19	54.8	2.05
2.83	70.7	5.7
3.35	83.7	11.2
3.80	94.9	18.6
4.00 (edge)	100.0	22.8

the curve flattens. Now if we pick several spots along the line between the center of the mirror and its edge, each locality may be treated as though it were an arc of a circle, with each arc having a longer radius of curvature than the one immediately preceding it. We shall measure these radii of curvature with the Foucault test apparatus, using another formula which resembles one we have used before.

In chapter 2 we worked out a depth of curve for the mirror which was in accordance with the focal length we wished to produce, using the formula $r^2/2R$, where again r stands for the radius of the mirror and R its radius of curvature. But this formula also gives the knife edge movement for a paraboloid if knife edge and lamp source move as one unit. In testing now, we shall keep the lamp stationary and move only the knife edge. This will double the distance the knife edge must move, so our formula becomes $r^2/2R \times 2$, or r^2/R. This formula will tell us the distance *the knife edge must move* from the position where it indicates the center of curvature of the mirror to measure the radius of curvature of any other zone on the mirror.

Now if we consider the mirror as a *series* of mirrors, each having a radius limited by the zone we are testing, we can make up another table, based on the first, which will give us the relationship of the position of the knife edge to the departure of the mirror from the sphere.

There are several things about the table on the next page which are worth looking at a little more closely.

1. The difference in positions of the knife edge, in testing zones from the center to the edge of the mirror, is only about $\frac{1}{7}$ inch. This is why we have to make the Barr scale so accurately.
2. Ten per cent of this distance tests a zone whose radius is about $1\frac{1}{4}$ inches, or 32 per cent of the distance from center to edge.
3. Fifty per cent of the total knife-edge difference in positions tests a zone whose radius is 2.8 inches, or 70 per cent of the distance from center to edge of the mirror.

Radius of Mirror		Change in Position of Knife Edge (r^2/R) from Center of Curvature		Actual Deviation from Circle $(r^4/8R^3)$ (in inches)
(in inches from center)	(per cent of total radius)	(in inches)	(per cent of total distance of knife edge)	
0	0	0	0	0
1.26	31.6	.0143	10	.224 $\times$ 10^{-6}
2.19	54.8	.0429	30	2.05 ″
2.83	70.7	.0715	50	5.7 ″
3.35	83.7	.1011	70	11.2 ″
3.80	94.9	.1289	90	18.6 ″
4.00	100.0	.1430	100	22.8 ″

4. The two measurements shown above indicate very clearly that the zone positions move ahead of the knife edge, a greater amount near the center but less and less as the edge is approached.

5. Since the zones near the edge have a greater radius of curvature than those at the center, the motion of the knife edge must be *toward* the observer to find the radius of curvature at the edge. In other words, if the mirror is a paraboloid, the central zone will darken when the knife edge is at zero position but the edge zone will not darken until the knife edge has been moved .143 inch toward the observer. But this tells us only that the center and edge of the mirror have the difference in knife-edge measurements that we expect for a paraboloid. In order to make sure that the intermediate zones are following the paraboloid curve, we must test them also. If the knife-edge settings in column 3 cause the proper shadows to appear in the positions indicated in columns 1 and 2, the mirror has the proper curve from edge to edge, and, in the absence of other irregularities, will give us the highest possible performance.

THE RAYLEIGH STANDARD

Irregularities on a mirror surface, either localized ones or the distortion of the paraboloid into some other figure, cause the reflected rays of a point of light to be shifted from their proper paths. Lord Rayleigh, the distinguished English physicist, showed that if the paths of any two light rays reflected from a mirror to a focus do not differ by more than a quarter wave length of light, the mirror will behave as though it were perfect. In other words, since the expanding cone of light from a star is made up of rays which strike the surface at an infinite number of points, each point on the mirror must be perfect enough to reflect the rays back to a focus in which each ray will be separated from any other ray by a distance not greater than a quarter wave length of light. But what does this represent in concrete figures?

Ordinary light is a mixture of rays of various colors, each having its own wave length. A wave length is considered to be the distance between the crests of the waves as the light travels through space. Red light has the longest wave length of the components of ordinary light and violet the shortest. The standard used is one about midway between these two extremes, or yellow-green, which has a wave length of 0.000022 inch. Now we are talking in terms of actual measurements; if we are to make a perfect mirror, we must smooth its surface to the extent where its reflected rays will come to a focus within an area whose diameter is ¼ of the wave length of yellow-green light, or 55 ten-millionths of an inch!

Nor are our troubles ended. Because of the laws of reflection, an error in curvature on the surface of a mirror will double the error in reflection of any given ray. This means that the errors on the mirror's surface must be *half* the tolerance we allow for the error in the paths of the light rays. In short, the Rayleigh Criterion, when applied to the mirror's surface, requires that the curve be adjusted to within ⅛ wave length of light, or, as it is usually expressed, ⅛λ.

This is theory. Lest you be discouraged by the infinitesimal size of the figures given above, you have already constructed an apparatus which will measure deviations of this order (your Foucault tester) and the techniques you will use for parabolizing will enable you to alter the surface of the mirror to compensate for such deviations. Furthermore, the Rayleigh limit may even be doubled and you will still have a mirror that will perform brilliantly. For this reason, you can permit deviations on your first mirror which do not exceed ¼λ, or .0000055 inch.

Now let's get down to cases. What does the theory given above have to do with actual practice? Referring to the table again, we see that the greatest deviation of the paraboloid from the sphere for a mirror of our diameter and focal length is .0000228 inch at the edge, and by lesser amounts as we approach the center. Therefore, we must transform the spherical surface by removing glass, in the parabolizing to be done in the next chapter, to depths corresponding to these figures, and we shall have a paraboloid. This would represent something close to perfection in the paraboloid, but we don't need to do this well, for the adjusted Rayleigh limit allows us room for error. Let's take the figure for the outside zone, .0000228 inch. We can depart .0000055 inch on either side of this distance, or we are allowed a variation of 25 per cent. This variation can be applied to all the other zones as well.

Now, since deviations on the mirror's surface are accurately reflected in the knife-edge positions which test them, we may also permit this per cent of variation in the knife-edge readings. As an example, the difference in knife-edge settings between the radius of curvature of the mirror's center and that of the edge is 0.143 inch. If the readings may vary 25 per cent of this distance either way, we may be satisfied with any reading for the edge zone which lies between 0.107 inch and 0.179 inch. Actually we shall try to do better than this, for we should make it a principle always to work as close to theoretical values as is humanly possible. No mirror maker worth his salt ever uses any permissible tolerance as an excuse for sloppy work.

At this point there is considerable variation in the practice of mirror making. There are those who say mirrors must be corrected *less* than these theoretical values since temperature effects encountered when the mirror is in use tend to deepen the curve. Telescopes are used mostly in the early evening when the temperature is falling and the mirror is contracting. During this time the colder external areas of the mirror exert a squeezing action which deepens the curve. After the whole mirror has cooled to a uniform temperature it resumes its normal shape, so that a mirror properly corrected for 75° above zero will have the same figure as at 10° below zero. But we are using it during the contracting period and therefore must expect a deeper curve than we actually put on it in the workshop. For this reason, some workers correct a mirror to only 95 per cent of the theoretical value, expecting that changing temperatures will do the rest. This is not necessary with Pyrex mirrors, since their coefficient of expansion is so low that the effect of temperature changes is almost negligible. So correct your mirror all the way and let temperature effects do what they will. The theory here is that the wise astronomer does not use his telescope during a period when the thermometer is jumping around, but uses it at as near a constant temperature as possible, secure in the knowledge that a properly corrected mirror will give him the best results.

The word "correction," as you may have gathered from the above, means the extent to which the curve has been deepened toward the paraboloid. A *fully corrected* mirror is one which has a parabolic curve. *Overcorrection* means that the curve is deeper than the paraboloid, becoming a hyperboloid. *Undercorrection* involves a shallower curve, which is a prolate spheroid or ellipsoid and not a paraboloid.

All this talk of accuracy in parabolizing may seem unduly fussy to those who have read elsewhere that one doesn't even need to parabolize a mirror and that, for most purposes, it may be left spherical. It is true that for long-focus mirrors, f/12 or over, the difference between sphere and paraboloid is very small indeed, and these mirrors may well be left spherical. But for an f/7, the departure between sphere and paraboloid is very appreciable. Therefore parabolizing to the best possible accuracy is essential. If the telescope is used for more exacting purposes—picking up the fine detail on a planet, for example—the demands made on the mirror surface become more rigorous. The requirement of ¼ wave length is not too exacting as far as the effort to produce it is concerned, and half the fun of making a mirror is the beautiful precision you can accomplish. It may help you to know that advanced amateurs usually work to within ⅒ wave length.

THE PARABOLOIDAL SHADOWS

Before you start to transform the polished spherical surface of your mirror into one that is parabolized, you must learn to recognize the shadow pattern characteristic of the paraboloid. If you raid the kitchen for a doughnut, lay the borrowed pastry flat on a piece of paper, and shine a light across it at an angle, you will have an approximation of what this shadow looks like. The difficulty is that all the figures of revolution except the sphere present some variation of doughnut shadows, and the only way you can determine which is the desired paraboloid is to make some measurements of position and intensity of the shadow pattern.

So there are three tests that you must apply before you can be sure that your mirror has the curve you wish.

1. Obtain some sort of doughnut shadow on the mirror.
2. Test the position and intensity of the shadows.
3. From observation of the mirror as a whole, be sure that the curve is a *smooth* one.

There are two methods of making these tests. A mirror which passes with a high score by either method is parabolized. We discuss both methods because of individual preferences in the mirror-making fraternity. There are devoted adherents of each method; we suggest that you read the procedure for each, then choose the one that you find the easier to understand and use.

THE COUDER SCREEN

This is a device worked out by André Couder, head of the Paris Observatory. It consists of a perforated screen placed over the mirror, the perforations being placed at the points where particular shadow patterns are to be observed. These holes in the screen are of different shapes to accommodate the various shapes of the shadows which they expose. Any number of zones may be tested, and appropriate holes cut for them, but for practical purposes only four are needed. If the mirror tests right for these four zones and the areas between them show a smooth, even surface, all intermediate zones must be correct also. As indicated in the diagram, the center of each perforation for an f/7, 8-inch mirror will be located as follows:

1. At the center of the mirror
2. 1.26 inches from the center of the mirror, or 32 per cent of the radius
3. 2.83 inches from the center of the mirror, or 70 per cent of the radius
4. 3.8 inches from the center of the mirror, or 95 per cent of the radius

The three outside pairs of holes should be at least ½ inch wide and 1½ inches long, while the center hole may be twice as large (see diagram).

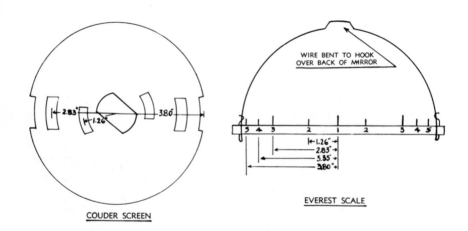

Two devices for measuring shadows.

(The figures are for the zones on an 8-inch mirror.)

TESTING WITH THE COUDER SCREEN

1. Testing with a Couder screen or with the Everest scale (which we discuss below) should be done in a faintly illuminated room, not in darkness. Faint differences in shadows are most readily seen in this way.

2. Place the mirror in its testing rack. Be sure that it is facing the slit in the Foucault lamp, as before. With the knife edge completely out of the cone of light so that the only illumination falling on the mirror comes from whatever room lighting you have, check the various open-

ings in the screen for even lighting. They should all be the same. Now remove the screen.

3. Adjust the slit in the lamp to 10–20 microns (3–6 pieces of typewriter paper) and line up the knife edge so it is parallel with the slit. This is done by pushing the knife edge an inch inside focus and watching the shadow on the mirror; it must be vertical. When your head is bobbed up and down behind the knife edge the line on the mirror must remain stationary. If it has a tendency to move slightly from side to side, the knife edge and slit are not parallel. If this is the case, adjust the knife edge in its holder until no lateral motion can be seen.

4. Adjust the knife-edge carriage so that when the knife edge penetrates the beam it may be moved back and forth without changing the amount of penetration. In other words, you should be able to push the knife edge to positions inside and outside of focus without causing the mirror to darken all over or become brightly illuminated.

5. Replace the screen and move the knife edge to the point where the center of the mirror darkens evenly; then move it back (toward you) to the point where the edge zone also darkens evenly. Be sure each of these positions lies on the Barr scale or within the limits of whatever measuring device you are using.

6. Now move the knife edge back to the position where the central hole darkened evenly all over as the knife edge cut the beam. Don't push the knife edge all the way across the beam in an attempt to make the mirror become completely black. The best results are obtained when the knife edge penetrates only the *edge* of the beam. At this point the central hole will appear a uniform gray color and minor gradations of shading will show up much more clearly than if the mirror is completely dark. You will be troubled by diffraction effects at the edge of the hole; try not to look at the bright rim of light produced here but concentrate your attention on the hole itself. When you have decided that the knife edge is at the exact point you wish, read the scale. This will be the zero point of your measurements and you will base all subsequent readings on this one. To make the description a little more real, let us make some imaginary readings and later on figure out what they mean as far as the mirror is concerned. Suppose, then, this first reading is .473 inch.

7. Now move the knife edge back until the second zone darkens. This is more difficult to evaluate since you are now trying to compare light intensities inside two holes instead of one. Don't try to look at both areas at the same time in order to compare them. Instead, let the eyes flick back and forth until you find a knife-edge position where, at the beginning of penetration, each hole seems to be equally gray. Again ignore the bright diffraction edges of the holes. Again, let us suppose that the Barr scale shows a reading of .468 inch, which means that the difference of radius of curvature of the two zones is .005 inch.

8. Move the knife edge back and test the third zone. These holes are even wider apart and your difficulties will be correspondingly increased. But keep trying until you are convinced that these holes, too, show an equal amount of grayness. This is a key zone and you must be especially careful here. Your reading, at this point, let us assume to be .399 inch.

9. Finally, move the knife edge back until the edge zones darken uniformly. Here you will be bothered by diffraction effects not only at the edge of the hole in the Couder screen but also on the edge of the mirror itself. Patience and perseverance will finally give you an accurate reading which, for this zone, we will assume to be .360 inch.

10. Now let us tabulate the results. The base, or zero, position was .473 inch. To find the relative distances of the knife edge from position to position, subtract each reading from the zero position, as follows:

1	2	3	4
.473	.473	.473	.473
.473	.468	.399	.360
.000	.005	.074	.113

Comparing these readings with the theoretical positions of the knife edge from zone to zone (see table on page 100), we have:

	1	2	3	4
Actual	.000	.005	.074	.113
Theoretical	.000	.014	.072	.129
Error	—	64.2%	2.7%	12.4%

These results, except in the second zone, are well within the permissible 25 per cent of error. The readings, in this hypothetical case, are quite close to those we hoped for, and satisfy the tolerances of the Rayleigh limit. However, to check against human error, take another set of readings and check them against the first. If they are quite close, use the average of the two sets for your final calculations. But if there is a wide variation, take a third set to determine which of the first two you wish to put your faith in.

11. Let us remember, however, that we have been looking at only a part of the surface of the mirror, that which can be seen through the holes in the Couder screen. It is quite possible that the measurements taken will satisfy the Rayleigh criterion, yet we may not have an acceptable mirror. In order to be certain that *all* parts of the mirror are as good as the sections you have tested, you must now check the surface as a whole.

Before continuing, let us look at one of the sources of confusion for many beginners. This is the similarity of terms used in discussing two related but independent measurements: the position of the knife edge and the position of the shadows on the mirror. Both are given as percentages. To avoid this confusion, it may be helpful to review some fundamental concepts.

a. Any given position of the knife edge, between settings for the center-zone radius of curvature and that of the edge zone, always produces definite shadow patterns on the mirror. These shadow patterns may represent imperfections on the surface or they may be the indication of definite mirror figures or curves. If the former, they indicate further steps to be taken in the process of parabolization.

b. The various knife-edge settings are always referred to as *per cent positions.* Thus the 10 per cent position of the knife edge means a point which is 10 per cent of the distance between its extreme settings, and this distance is always measured from the setting which measures the radius of curvature of the center of the mirror.

c. Zonal areas always refer to the mirror itself, not the knife-edge settings. Thus the 70 per cent *zone* is an area on the mirror which is $\frac{7}{10}$ of the distance from the center of the mirror to its edge.

d. The percentages spoken of for knife-edge positions do not correspond numerically to mirror zones. A 50 per cent knife-edge setting is related to the 70 per cent mirror zone. The correspondence of knife-edge settings to mirror zones is shown in the table on page 100. This correspondence may seem obvious to some readers; we speak of it because it has been a source of confusion to others.

e. When you are using the Couder screen, the knife-edge settings measure particular zones and give information only about those zones, but when the whole shadow pattern is looked at without the screen, *any* knife-edge setting gives information about the *whole* mirror. If the shadow position and motion are correct for a particular knife-edge setting, they will be correct for all the others (but we check them just the same!).

f. Always think of the shadow patterns as representing a cross-section of the mirror. As you become accustomed to the appearance of the shadows you begin to think in terms of high and low points which you can transform into a sketch of the cross section. The mirror is then treated according to what the cross-sectional sketch looks like; high spots to be planed down, low spots to be blended into the surrounding area, etc.

TESTING THE OVER-ALL FIGURE

Testing the Edge

The test made with the Couder screen on the No. 4 zone was not that of the outer edge, but rather that of a zone whose center has a radius of 3.8 inches. It is impossible to test the edge itself because of the diffraction ring which surrounds it and because we must test a fairly wide zone, not just a line. But if we move the knife edge back to the theoretical point which calls for maximum separation of sphere and paraboloid and observe the general position of the shadows, we are in effect testing the edge itself. If these shadows are in the right positions, the edge *must* be right.

In talking about the general shadow patterns on a mirror, we must define the two terms *crest* and *shadow* itself. By crest we mean the highest point in any apparent curve which is convex toward us; in short, the high point of any bulge. But when we speak of the shadow itself we mean the *edge* of the shadow, or that point where the fine spidery lines of light can no longer be seen. Furthermore, we must recognize that the intensity of the shadows varies greatly with the focal length of the mirror. For short focal lengths, up to f/6, shadows are dark and well defined. But for longer focal lengths, f/7 to f/10, the shadows become progressively more delicate and harder to see, and for mirrors of f/12 and over, no shadows are seen at all. This is because, as we said earlier, the difference between sphere and paraboloid for long-focus mirrors is minute. Mirrors over f/12 may be safely left spherical and in this condition are considered to be amply corrected by most mirror makers.

In testing for over-all figure by shadow patterns, we move the knife edge all the way into the cone of light, not stopping at the edge as we did when we used the Couder screen. We start by setting the mirror in its support, finding the zero position of the knife edge for its center, and then moving the knife-edge carriage back to the position where, when the Couder screen was in place, the edge of the mirror darkened evenly. This is the position for the radius of curvature of the 95 per cent zone.

Move the knife edge into the cone of light and watch the shadows as the penetration increases. At the beginning, a faint shadow should appear on the right of the mirror at the 70 per cent zone (radius 2.83 inches). Moving the knife edge farther in should cause this shadow to grow larger, expanding more rapidly to the left than to the right. It should reach the edge of the mirror on the right at the same moment that its other side reaches the center. At this instant a faint shadow will appear on the left of the mirror. Continue the lateral motion of the knife edge into the cone of light and note the relative motion of the two shadows. The one at the center moves more rapidly than the one at the left, but as the two approach each other they will meet at the 70 per cent zone on the left of the mirror. At this point the mirror is completely dark. If the shadows move according to this pattern, you may be sure that the mirror is a good one. It would be very difficult to make drawings to show all the minute changes which occur on the mirror's surface during the shadow transitions, so we shall show, in the drawings here, how

SHADOW PATTERN

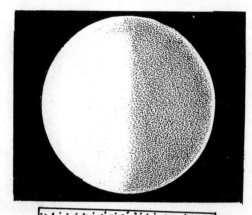

RAYS REFLECTED
FROM MIRROR

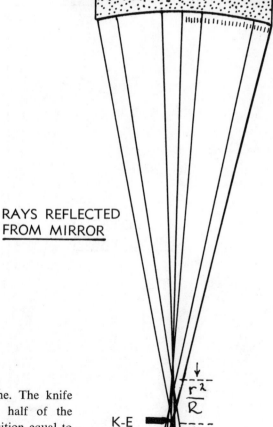

K-E

$$\frac{r^2}{R}$$

Testing the edge zone. The knife
edge has penetrated half of the
cone of light at a position equal to
90 per cent of r^2/R. If the shadow
pattern is similar to the one shown,
the mirror is paraboloidal.

SHADOW PATTERN

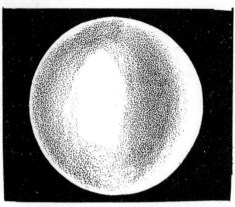

RAYS REFLECTED
FROM MIRROR

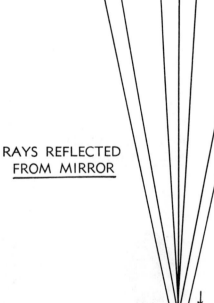

K-E

$$\frac{r^2}{R}$$

Testing the 70 per cent zone. The
knife edge is at a position equal to
50 per cent of r^2/R. The shadows
should appear as shown above.

APPARENT CROSS SECTION

SHADOW PATTERN

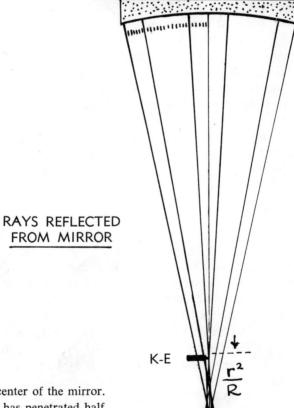

RAYS REFLECTED
FROM MIRROR

K-E

$$\frac{r^2}{R}$$

Testing at the center of the mirror. The knife edge has penetrated half the cone of light at a position equal to 10 per cent of r^2/R. The mirror should have the inverted bowl appearance shown above.

the mirror looks as the knife edge cuts the center of the beam of light. The second set of drawings shows what rays are cut by the knife edge, and the third set is an apparent cross section of the mirror.

Testing the 70 Per Cent Zone

With the knife edge in the 50 per cent position (halfway between testing positions for center and edge), the motion and position of the shadows give an indication of the mirror's condition in the 70 per cent zone, or 2.83 inches from its center.

As the knife edge begins to cut the cone of light, a shadow appears on the left edge of the mirror, moving toward the center. Almost simultaneously, a second shadow can be seen at the 50 per cent zone on the right, or halfway between center and edge. This second shadow expands as the knife edge cuts in farther, moving more quickly toward the right than toward the left. The first shadow, in the meantime, moves toward the center. When the knife edge progresses to the center of the cone of light, the appearance of the mirror, the rays cut, and the apparent cross section should be as shown here. Note that the first shadow has reached the 70 per cent zone on the left while the second shadow extends from the center of the mirror to the 70 per cent zone on the right. The apparent depth of the mirror should be the same at the center as at the edges.

Now continue the motion of the knife edge and watch the two shadows approach each other. They should meet at the 50 per cent zone on the left. At this time the right-hand shadow should have not quite reached the edge of the mirror.

Testing at the Center

Move the knife edge forward to the zero position. As you cut the cone of light a shadow appears on the left edge of the mirror. It moves as the knife edge progresses, slowly at first, then more rapidly until, when it has reached the 50 per cent zone, it suddenly sweeps across the center of the mirror. The knife edge has now reached the center of the cone of light, and the mirror looks like the base of a flat bowl, seen bottom side up. The diagram here shows the general appearance, the rays cut, and the apparent cross section. Move the knife edge back and forth in the beam—a *very* slight motion is all that is needed—watching the upper and lower shadow boundaries while doing so. They should appear to pivot on the top and bottom poles of the mirror, and there should be little or no distortion of the shadow edge. If even a small distortion is noted, move the knife edge ½ inch inside the focus and again bring the shadow in from the left-hand edge. With the knife edge in this position, the shadow should have a slight curve inward with no change in direction as it runs off the edge. If it bends farther in, the mirror has a turned-down edge. But if it straightens or bends outward, the edge is turned up.

Testing for Astigmatism

You have now made two series of tests. If your mirror has passed both series, it is almost a certainty that it will perform brilliantly when mounted in its tube. There is still a faint possibility, though, that the mirror may be astigmatic, which means that the same curve is not present on every diameter of the mirror. You can check this by watching the shadow move across the mirror as you push the knife edge through the cone of light at a point an inch inside or outside of focus. The shadow must travel horizontally; it must enter on the extreme left of the mirror and leave on the extreme right. If it gives any indication of coming into or leaving the field in any other than a horizontal direction, the mirror is astigmatic. Test three different diameters in this way by turning the mirror through 60° after each test.

THE EVEREST TEST METHOD

The Couder screen test and its variations—actually this test was originated in a slightly different form by the great mirror maker Ritchey —is the classic test used by many amateurs, from beginner to expert. It has two disadvantages, however. The first and most bothersome is the presence of diffraction rings on the edges of the holes in the screen. These are very distracting when you are trying to decide on the intensity of light seen inside the holes on two different parts of the mirror. The second disadvantage is that the screen must be removed to test the over-all surface of the mirror, with the inevitable consequence of disturbing the whole setup and having to rearrange it.

A simpler and equally effective method of testing is the one invented by A. W. Everest.[2] In this method the screen is dispensed with and the shadows are located by pins set in a thin wooden strip placed across the face of the mirror. The zones to be tested are decided upon first, and their radii worked out in the same way as in the Couder screen method. Since there are no holes to cut, there is no problem of spacing them at odd places to make room for all the holes (and therefore zones) we might wish to test. So we can test five zones easily, and will adjust the test apparatus accordingly.

Make a $9 \times \frac{1}{4} \times \frac{1}{4}$-inch stick from a piece of straight-grained pine or spruce. Attach a wire loop to each end of the stick, adjusted so that when the loop is hung over the top of the mirror, the stick will be supported

[2] Albert G. Ingalls (editor), *Amateur Telescope Making,* Book II, Scientific American Publishing Company, New York, pages 21-26.

horizontally with its upper edge at the mirror's center. Place a common pin in the center of the stick; then place other pins at distances of 1.26, 2.83, 3.35, and 3.80 inches on either side of the central pin. Blacken these pins so they will not reflect too much light. They will be easily visible in the testing procedure because of the slight diffraction effects they cause.

The actual testing is similar to the procedure followed for the over-all testing of the mirror, but there are enough differences so we shall describe it in detail. It is just the reverse of that done when the screen is used. Instead of finding shadows of equal intensity on the zones of the mirror and then locating the corresponding position of the knife edge, we place the knife edge in each of its five positions and look for shadows at the pins which correspond to these positions. This may not seem like an essential difference, but it makes testing much easier because, if misplaced shadows show up, we know their actual location on the mirror itself instead of having to translate differences of knife-edge readings into corresponding defects.

The Testing Procedure

Locate the 95 per cent zone on the mirror as described earlier (page 108). This will give you a scale reading for the 90 per cent position of the knife edge, and from it you can estimate the zero position. The other positions of the knife edge will be measured from this starting point as follows:

Position of Knife Edge	*Zone on Mirror*
1. zero	Center of mirror. Pin 1
2. 10% position, .014″ from zero	31.6% of radius, 1.26″ from center. Pin 2
3. 50% position, .072″ from zero	70.7% of radius, 2.83″ from center. Pin 3
4. 70% position, .101″ from zero	83.7% of radius, 3.35″ from center. Pin 4
5. 90% position, .129″ from zero	94.9% of radius, 3.80″ from center. Pin 5

1. Place the knife edge at the 90 per cent position. The crest of the shadow coming in from the left of the mirror and the crest of the shadow extending to the right from the center of the mirror should each fall at the 95 per cent pins (pin 5). The edge of the shadow should be at 70 per cent of the 95 per cent zone, or 3.61 inches from the center. There is no pin at this point, but the position can be estimated with sufficient accuracy.

2. Move the knife edge to the 50 per cent position. Check the position of the crests of the left-hand and right-hand shadows to see whether they fall on the 70 per cent pins (pin 3). Note also whether the shadows reach the pins at the same time as the knife edge cuts the cone of light, and also

whether the apparent depth of the mirror is the same at the center as at the edge. The edge of the left-hand shadow should be halfway from center to rim.

3. Move the knife edge to the 10 per cent position. This is a difficult zone to test since the gradation of shadow intensity is slight. But you should see just a faint evidence of an apparent depression at the center of the mirror, rising to a crest at the 32 per cent zone (pin 2). The motion of the shadows will also be difficult to detect, and it will be hard to say whether or not the shadows reach their respective pins at the same time. Careful control of the knife edge as it is inserted into the beam of light and repeated trials will enable you to distinguish the shadows and their motions, so that you will become as confident in testing this zone as in testing any of the others.

4. Move the knife edge to the zero position. The central zone should now appear nearly flat (pin 1). The whole mirror, when the knife edge cuts the center of the zone, should have the inverted bowl appearance described earlier (page 112).

5. Move the knife edge to the 70 per cent position. Here the left-hand shadow enters the left edge at the same time that the right-hand shadow is moving away from the central zone of the mirror (pin 4). And, as it progresses across the mirror, it should also enter the central zone at the moment that the right-hand shadow is passing off the edge of the mirror.

Testing these five zones gives us all the information we need about the mirror. We could test more than five, of course, in the hope of getting all the information we can squeeze from our test apparatus. But since we shall test often during the parabolizing period, time now becomes a factor and it is therefore desirable to hold the procedure down to the minimum number of steps. Some authorities suggest testing only three zones, the center, edge, and 70 per cent areas, based on the assumption that the mirror will be parabolized if these zones check out properly and if a smooth curve lies between them. This is an erroneous assumption, since *any* of the figures of revolution (except the sphere) will pass this test. The mirror could be badly overcorrected (too near the hyperboloid) or undercorrected (too near the oblate spheroid) and it would pass unnoticed if only these three zones are checked. Testing at least one other zone is therefore essential. Two more will tell us the whole story about the mirror's condition from edge to edge.

CHAPTER 9

Parabolizing the Mirror

WORKING CONDITIONS

Eᴀʀʟɪᴇʀ ɪɴ ᴛʜᴇ ʙᴏᴏᴋ it was pointed out that mirror making must take place under certain fixed conditions. The temperature of the air, the temperature of tool and mirror, the humidity, and the presence of stray air currents must be carefully controlled. Important in any stage of mirror making, these conditions become vital during the final stages of testing and figuring. Even a slight difference in temperature between tests will change the figure on the mirror, low humidity produces rapid evaporation from lap and mirror and thus causes uneven cooling of surfaces, air currents can produce weird effects in what is seen on the testing stand (imagine the wavering currents of air which you see on a warm summer's day magnified a hundred thousand times!).

Air currents can be controlled by using a testing tunnel made of plastic on a wooden framework. The framework is made just large enough so that the testing stand can be introduced through a port at one end. The Foucault tester is placed at the other. Devices such as this, although troublesome to make and maintain, are invaluable if you are not fortunate enough to have a small constant-temperature room in which to work. If you lack such ideal facilities, minimum precautions to ensure consistent test results are (1) to hang a thermometer somewhere in your workshop and test only when the thermometer registers a predetermined constant reading, and (2) to keep the floor damp or place flat pans of water at strategic points to control humidity as much as possible.

Other essential precautions include not handling the mirror any more than you can help after it has been placed on the testing stand, reducing polishing periods to the point where little heat can be built up on mirror and lap, and allowing the mirror to cool for at least 20 minutes before testing. For final tests, wait from 1 to 3 hours.

INTERPRETATION OF TEST OBSERVATIONS

We used up much space in the previous chapter to describe various methods of testing the mirror for the paraboloidal figure. In this one we shall talk about the principal figures we may find on the mirror, the means of recognizing them, and ways of reducing each to the hoped-for paraboloid.

If you were very fortunate, you ended polishing operations with a spherical surface on the mirror; you might even have produced the paraboloid itself. In the event of either of these happy conclusions, work on the mirror is very nearly finished. But there is a possibility that the figure turned out to be either a hyperboloid or an oblate spheroid. An important question now is, how can you recognize either of these figures?

The best place to start is with the zonal measurements of the center and the edge. If the difference between these measurements is more than 25 per cent greater than the calculated amount—in this case, .143 inch— the mirror has become too deep and has a hyperboloidal curve. On the other hand, if it is more than 25 per cent smaller than this figure, the mirror is too shallow and probably has the shape of an oblate spheroid. Consequently, your first job is to reduce either of these curves to the point where it can be transformed into a paraboloid. To be perfectly safe, you should work toward a spherical surface which you can then deepen into a paraboloid. But a perfect spherical surface is as hard to produce as a perfect paraboloid, so you compromise on working *toward* the sphere, knowing that at some point or other you will arrive at a figure you can transform into the paraboloid.

CORRECTING THE HYPERBOLOID

Let us assume that your measurements indicate the mirror is hyperboloidal. You can confirm this suspicion by the appearance of the mirror as well as by the measurements between zones. Locate the 50 per cent position of the knife edge by halving the distance between center and edge measurements and cut into the beam of light. This position of the knife edge is a measure of the shadows on the 70 per cent zone of the mirror. The shadows you observe will be sharply defined and dark. The crests will be too far in on the right and too far out on the left. The shadows themselves will be narrow. The limits for the shadow on the right will be too near the crest, while the left-hand shadow limit will be too far from the crest. The center will appear to be deeper than the edges.

Such a figure is probably the result of long strokes in the final stages of polishing, and the obvious cure is to shorten the strokes in an effort to plane off the crests and make the curve more shallow.

Set the knife edge at a position to point out the areas where you will have to do the least amount of work to accomplish this planing action. This knife-edge setting should be at the point where the apparent depth of edge and center seems to be about the same. In practice you will find it quite close to the radius of curvature of the edge zone.

Use strokes which will bring the edge of the lap not quite to the crest, and polish for 2 or 3 minutes, using moderate pressure. Allow the mirror to cool, and test. The crests will appear to move out and the shadow limits widen. Cold press for 5 to 10 minutes, then continue the process until, at the 50 per cent position of the knife edge, the crests are in the right position as explained in the section on the Everest test method. In the meantime, if the diffraction ring at the left of the mirror becomes less bright, lengthen the strokes until it is brought back. This may seem a contradiction to the ordinary treatment for turned edge, but in this case the defect is caused by the plowing effect of the mirror edge on the lap. Here, lengthening instead of shortening the strokes is an effective measure.

This means of correcting the hyperboloid is difficult to control because of the short strokes, and it is quite possible that the result may be an oblate spheroid with zonal errors. This is no cause for worry, however, since the oblate spheroid responds to treatment readily.

CORRECTING THE OBLATE SPHEROID

The oblate spheroid (the figure in which the central zone is not deep enough in relation to the edge zones) can be corrected in much the same manner as removing a central hill. The figure is easily recognized for, at the 50 per cent position of the knife edge, it appears to have a flat center zone with upturned edges. The shadows are concentrated on the right edge of the mirror. When the knife edge is placed at the center of curvature of the edge zone, or in the 90 per cent position, the mirror appears to be flat on the edges and has a large central hill.

Set the knife edge where there is the smallest difference in apparent depth of zones, probably somewhere inside the 90 per cent position. Use a "W" stroke of ½ to ¾ the mirror diameter, with enough side overhang so that the edge of the tool moves in to the boundaries of the hill on each side. This stroke should be used for only 2 minutes, then check the mirror for the appearance of a hole in the center of the hill. If this happens, reduce the length of the stroke. But if the hill seems to be disappearing, subsequent periods of polishing can be increased in proportion to the effect the polishing action has on the hill. In order to maintain a smooth surface on the mirror, vary the strokes by using the blended overhang described on page 90. During the process, watch out for the appearance of a turned edge. A hard lap and frequent cold pressing will protect the edge, but if the diffraction ring indicates that the edge is turning, shorten the strokes. Also, as the hill disappears, keep an eye on the edge for a lengthening of the radius of this zone. If this happens, the strokes are too long. When the central hill can barely be seen, cold press for a half hour, then change to $\frac{1}{3}$ "W" strokes to remove any remaining zonal irregularities. The mirror should need only a few short spells of this to become spherical.

PARABOLIZING METHODS

If you were careful in the final stages of polishing, your mirror probably needed neither of the treatments suggested above. It is more likely that you ended polishing with a mirror that was spherical, or nearly so. Now you can enter the final stage of mirror making, and if all goes well your work is nearly done.

There are several methods of parabolizing, any of which will reach the goal if properly followed. None is especially difficult, so it may be useful to present them all and let the reader take his choice.

Parabolizing by Long Strokes

We learned early in the grinding process that a long stroke will rapidly hollow out the center of a mirror. This is exactly what we wish to do here, but the hollow must be one of smoothly sloping sides which extend all the way to the edge of the mirror. Consequently the strokes used must not only do the job of excavation but also preserve the smoothness of the surface as a whole.

Start with a well-pressed lap, making sure that the mirror makes contact everywhere. It is a good idea to cold press with window screening to increase the facet area. Apply fine washed rouge more liberally than usual, and start by using very long strokes, as much as $\frac{3}{4}$ of the mirror diameter. These are used in "W" form, with liberal overhang on each side of the lap. Work slowly and carefully, polishing for $\frac{1}{2}$ a minute, then cold pressing for 2 minutes. It is essential in parabolizing that good contact be maintained at all times; this is the reason for the frequent cold pressing. After a total of 3 minutes of polishing time, remove the mirror, allow it to cool, and test. If a figure has started to appear, check the positions of the crests of the shadows with the knife edge in the 50 per cent position. If they are too close together, shorten the strokes; if too far apart, continue with the long ones. Keep on with the process, polishing, testing, and pressing, until the crests are in the right position. If the long strokes have produced zones in the mirror, remove them with the blended overhang. Many workers like to use elliptical strokes at this point. They produce excellent results if used carefully; otherwise they will quickly turn the edge. Concentrate any stroke used as though you were pushing glass from the center of the mirror toward the edges, with the edge of the tool doing most of the work.

When you are sure that the crests are in the right position for all zones, let the mirror sit for an hour, then test again. You may be surprised at the difference between the test results. But if they check with each other, carry out tests on all other zones. If the shadow patterns and knife-edge positions are in close agreement, stop here.

Parabolizing by Small Polisher

Some workers find it more convenient to parabolize a mirror using a small lap since it is easier to control the figure if the area to be worked on can be isolated. Place the mirror face up for this operation and make a ¾ diameter pitch lap. If you made a sub-diameter lap for any stage of

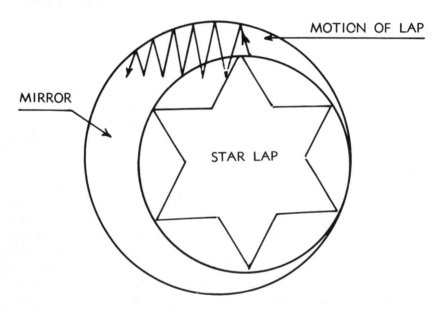

Using a sub-diameter lap.

the polishing process, you can use it here. Transform it into a star lap by cutting away the outside facets to form the points of a star. The star lap works very smoothly and its action is easy to control, but some workers find it difficult to make because of the tendency of the pitch to chip. But if you use a very sharp X-acto knife and work cautiously, this lap is as easy to make as any other kind.

After the lap is in good contact with the mirror, apply a fine washed rouge—be careful here, for all operations where the mirror is used face up are invitations to scratching—and start polishing. The edge of the lap

Star lap for use with an 8-inch mirror. Some of the facets were damaged in cutting away the pitch and must be built up again.

should just reach the edge of the mirror at the end of the stroke, and a series of strokes should cause the lap to travel across the mirror as shown in the diagram. Again the idea is to push glass away from the center of the mirror. This operation is a relatively slow process as far as the total effect on the mirror is concerned, but stop to test frequently anyway. Try a 5-minute period of polishing for the first spell, and judge the polishing periods thereafter by the effect the first one has on the mirror.

As soon as a figure appears, test from the 50 per cent knife-edge position and check the position of the crests in the 70 per cent zone. If the crest is too far out—which it probably will be on the first test—concentrate the strokes more on the center of the mirror as the strokes move from side to side. But if the crest is too far in, the marginal zones should get more strokes per inch of lateral motion. In all changes of strokes, however, remember that a small polisher has a drastic action on local areas, even though its total effect on the mirror may be slow. It can make a hole very quickly anywhere on the mirror. Hence you must concentrate on trying to produce a smooth, wide depression in the center of the mirror, not a small one which you will extend later. In other words, try for the parabola from the first appearance of the figure; then concentrate on deepening the parabola as you go along.

Parabolizing by Deformed Lap

The classic method of parabolizing is to depress the outside facets of the lap, either by cutting them away completely or using some means to push them down far enough to nullify their action on the mirror. Still another way is to trim their edges progressively from the center to reduce their effectiveness from center to edge. If all goes well, the mirror can be parabolized quickly in this way, using strokes of from $\frac{3}{4}$ to $\frac{4}{5}$ the mirror diameter. But in the event that a hyperbola is produced or that zonal irregularities appear on the surface, the worker is faced with the necessity of attempting to remove the defects by using small polishers or by making a new lap. Consequently few mirror makers are now willing to undertake this method, especially since there are others equally good but involving less risk.

Parabolizing by Reversed Lap

In this method the tool is placed on the face-up mirror and the parabolizing is accomplished by varying the position of the lap as it travels across the mirror and also by varying the pressure on the edge of the lap. In effect, the method is almost exactly like that of using a sub-diameter tool, the difference being that a part of the full-sized tool is used instead of making a smaller one. Since the edge of the tool exerts the main polishing force, the remainder of the tool must overhang the edge of the mirror to a considerable extent. The overhanging part is guided by one hand, exerting just enough upward pressure so the edge of the mirror is not turned, while the other hand exerts pressure on the inner edge which travels back and forth on the center of the mirror. Use the same technique as in other methods: make a concentrated effort to push glass away from the center of the mirror along an even slope.

The difficulties are increased since the tool deforms rapidly because of the localized pressure, and constant cold pressing is required to maintain good contact. Also it is difficult to judge how much pressure is to be used on the bearing edge and how little is required to support the overhanging edge. A mistake in judgment will result in a seriously turned edge very quickly. On the other hand, the method is very effective, once mastered, especially with mirrors of short focal length. Furthermore, by using the center of the lap on the edge of the mirror—here the pressure is distributed evenly—the method can be used to plane off raised zones near the edge.

This method typifies a fundamental requirement for all parabolizing efforts. It is the relationship between thought and effort. Thomas Edison once said something about genius being 90 per cent perspiration and 10 per cent inspiration. If this principle were applied to the final stages of mirror making, no mirror would ever be completed, for here the proportions of inspiration and perspiration must be just the reverse of Edison's figures. A little thought will save you hours of work. If you think of the elimination of zonal irregularities or of grosser errors of figure in terms of what caused them, the cause itself will point toward the cure.

ACCURACY IN PARABOLIZING

Earlier in the book we spoke of the Rayleigh criterion in connection with accuracy in parabolizing. We pointed out that we must produce a surface accurate to within $\frac{1}{4}$ wave length of light if we wish the mirror to perform at its best. But why stop here? Why not make a mirror with no greater deviation than, say, $\frac{1}{20}$ wave? And if we do so, how much better will it perform than if we are satisfied with $\frac{1}{4}$ wave?

Let us answer the last question first. In order to understand how a mirror produces an image we must understand something of light itself. Light travels in waves whose lengths vary roughly from .4 to .65 microns. The very term *wave* implies a vibration of the light. It would seem impossible that a wiggling ray of light could ever be focused into a sharp point; in other words we should not expect a light wave to concentrate its energies on a single point any more than we should expect an ocean wave to concentrate on a single pebble. And this is just what happens.

Let us think about a star so far away that, for all practical purposes, all its light comes from a single point. No matter how perfect an optical instrument we may possess, the light from this star cannot be focused into another single point. It will be concentrated in a disk instead. This is what we should expect from a bundle of light rays, each of which is wiggling in a different direction laterally. Furthermore, this disk will be surrounded by dark rings, where the waves interfere to the point that they cancel one another out—the crest of one wave falls on the hollow of another. Between the dark rings will be bright ones, where the waves reinforce one another, or crest falls upon crest. This bright spot of light and its attendant rings is called a diffraction pattern or disk, and is formed by each tiny area of the mirror reflecting rays back to the same spot. The more nearly perfect the surface of the mirror, the smaller and brighter this disk will be, and the better the image formed.

The size of the disk produced by a perfect optical system was first calculated by Sir George Airy in 1834, and is known as the Airy disk. He found that for a circular aperture focused perfectly, 84 per cent of the light will concentrate into a central disc, and the remaining 16 per cent into a system of surrounding rings. The disc is not a uniformly illuminated one, but is highly concentrated in the center, fading off to zero illumination at the edge and forming what is known as the first dark ring, the radius of which is:

$$r = 1.22\lambda f/D$$

where r = radius of the first dark ring
$\quad \lambda$ = wavelength of the light
$\quad f$ = focal length of the mirror
$\quad D$ = diameter of the mirror

Applying this formula to an f/7 mirror of 8-inch diameter, we would have

$$d = 1.22 \times .000022 \times 56/8 = .0002$$

Put into more concrete terms, this means that all rays coming from the surface of an f/7 will be focused in an area only 2 ten-thousanths of an inch in radius if the mirror is perfect. Deviations on the surface of a mirror will cause the rays to fall outside this area, with a consequent loss of resolving power.

If we consider the diameter of the diffraction disk from another viewpoint—for example, the size of the angle it subtends as seen from the mirror's surface—it can be shown that the angle equals $11/D$ seconds of arc, where D is the diameter of the mirror. From this formula it becomes evident that the larger the mirror the smaller the diffraction disk and the smaller the subtended angle. The idea of translating linear diameter into angular diameter (or the size of the subtended angle) may become more clear if we consider the following example.

A dime is about $1\frac{1}{16}$ inch in diameter. If you hold it at arm's length, the dime will just about cover the face of the full moon. Since the moon covers an arc of the sky 31 minutes wide, we can say that the dime, held in this way, has an angular diameter of 31 minutes. Similarly the diffraction disk of a perfect mirror of 8 inches diameter will have an angular diameter of $1\frac{1}{8}$, or 1.37 seconds of arc. (1 degree equals 60 minutes equals 3600 seconds.)

Now let us apply this concept to two equally bright stars, close together in the heavens. Each will produce a diffraction disk in the telescope. If these disks fall on top of one another, the two stars will appear as one. But if the disks are separated by half their diameter or more, the stars will appear as two interlocked entities. Thus we may reduce the above formula from $11/D$ to $5.5/D$ for the purpose of separating two close objects, and expect that the 8-inch mirror will separate objects whose angular distance in the heavens is only .68 seconds of arc.

This is theory; in actual fact the formula may be reduced still more, as discovered by the nineteenth century astronomer W. R. Dawes, who set the limit as $4.56/D$. The Dawes limit, as it is now called, predicts that an 8-inch mirror should separate two stars only .57 seconds of arc apart. A telescope which performs within this limit is a good one indeed, and it is the criterion which most amateurs use as a measure of the perfection of their instruments.

This is as good a place as any to distinguish two terms which are often confused in talking about mirrors: resolution, and definition. *Resolution,* or resolving power, is the ability of the mirror to separate close together objects such as double stars, and depends upon the aperture of the telescope. *Definition* is a measure of the fidelity of the images throughout the field. It is directly proportional to the focal length of the mirror as well as to the quality of its surface.

A consideration of the Dawes limit presupposes that the magnification of the telescope (see chapter 11) is such that the tiny angle can be enlarged to the point where the eye can observe it. Most of us have eyes which can distinguish objects separated by 4 minutes of arc, under ordinary lighting and without optical aid. The eyepiece of the telescope therefore must increase the angular diameter of the image formed by the mirror from .57 second to 4 minutes of arc, an increase of approximately 420 times. Now, observing conditions where magnification of this order can be used are very rare. Disturbing elements such as changing temperature, turbulence of the atmosphere, air currents within the telescope tube, etc. must be at a minimum. However, if all elements thus far mentioned—definition, resolving power, and observing conditions—are at a peak, the telescope will perform at the limit of its potential. Even so, we may still have performance far beyond that of ordinary commercial instruments even if the mirror falls somewhat short of the perfection mentioned above. The minimum requirement is $\frac{1}{4}$ wave, which will yield 40 per cent illumination in the diffraction disk. A perfect mirror gives 84 per cent illumination, while one corrected to $\frac{1}{8}$ wave produces 68 per cent. Let us set this intermediate figure as what we hope for in a first mirror, knowing that even if we fall considerably short we shall still have an instrument, all other parts of which are of equal quality, which will perform beautifully.

But how do we interpret the Foucault measurements to know when we have reached the $\frac{1}{8}$ wave requirement? Let us take the figures given on page 106 as a case in point. These show the actual measurements of a mirror as compared to what should be expected from theory. The third line of the table indicates the per cent error of the actual knife-edge readings, which may be translated into actual errors on the surface of the mirror, either mathematically or graphically.

The method which gives the clearer picture of the mirror condition is the latter, or graphic, method, in which the knife-edge readings are plotted against those for a true parabola as well as those for under-corrected and overcorrected mirrors. The limits of correction for an 8-inch mirror are 25 per cent over or the same amount under, or 125 per cent and 75 per cent correction, respectively. A mirror corrected within these limits will have no error greater than $\frac{1}{4}$ wave on its surface.

This information can be tabulated in the table at the top of the next page, which shows that the mirror is under-corrected in the first and last zones and slightly overcorrected in the intermediate one. But a graph shows this even more clearly, and furthermore

Knife-Edge Reading (inches)	Zone (in inches from the center)			
	Center	1.26	2.83	3.80
125% (overcorrection)	0	.018	.090	.161
100% (full correction)	0	.014	.072	.129
75% (undercorrection)	0	.011	.054	.097
Actual (see page 106)	0	.005	.074	.113

gives a rough quantitative estimate as to the deviations from full correction.

If we plot the values given above, using the vertical scale for knife-edge readings and the horizontal scale for mirror radius,[1] even the most cursory inspection will reveal the degree of overcorrection and undercorrection in the critical zones. If the space between the central line (the 100 per cent correction line) and either of the limiting lines indicates a quarter wave deviation from the paraboloid, the amount of overcorrection or undercorrection may be estimated roughly in terms of wave length. Furthermore, if a smooth line is drawn to represent the average of the points indicating the actual measurements, the average correction of the whole mirror may also be estimated. Obviously, this mirror needs treatment in the first and last zones, whereas the intermediate zone can be considered satisfactory.

The final question is, how accurate is the Foucault tester itself? The answer obviously depends on the accuracy used in its construction, the number of readings made in each zone, the conditions under which the tests are made, and the skill of the observer. Allan Mackintosh, using the Everest test method and the Mackintosh tester, can figure a mirror to within $\frac{1}{15}$ wave, results which later check out with more precise testing methods.[2] Most beginners, however, working by themselves, will do well if they exceed an accuracy of $\frac{1}{4}$ wave. Mirror making, like any other skill, is a matter of practice, as those who have made their second or third mirror can testify.

[1] See Albert G. Ingalls (editor), *Amateur Telescope Making,* Book I, Scientific American Publishing Company, New York, pages 257-261; and Jean Texereau, *How to Make a Telescope,* Dover Publications, Inc., New York, pages 93-100, for more detailed treatments of graphical analysis of a mirror.

[2] For a description of the so-called caustic test for mirrors, capable of detecting errors as small as 1/100 wave, see Albert G. Ingalls (editor), *Amateur Telescope Making,* Book III, Scientific American Publishing Company, New York, pages 429-456.

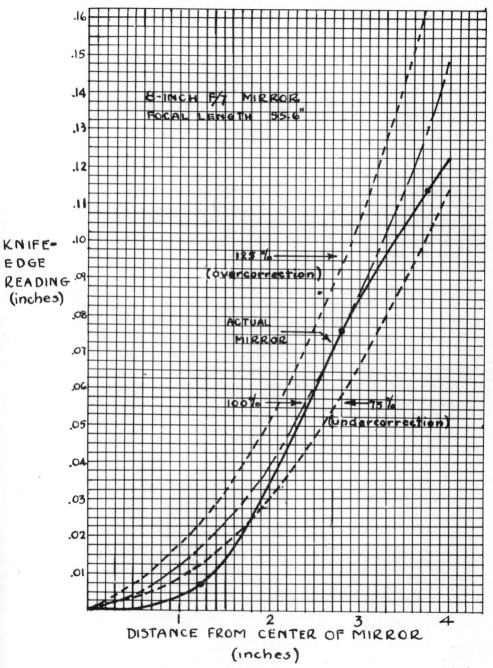

Graphical evaluation of a mirror.

The heavy line shows the mirror well corrected in the 70
per cent zone but undercorrected at the center and the edge.

MIRROR COATINGS

The piece of glass that is the end result of the polishing and grinding operations will not act as the mirror in the telescope; it is only the supporting structure for the real mirror, which is a thin metallic coating applied to the surface of the glass. Glass itself is not reflective enough to serve as a mirror; at its best it returns only 5 per cent of the light that strikes it. So the actual reflective surface will consist of a layer of silver or aluminum only a few molecules thick. It must be as thin as this if it is to reproduce the curve so carefully ground into the surface of the piece of glass. Consequently, if we expect it to cover up any defects which appear there, we shall be disappointed.

Up to a few years ago, most astronomical mirrors were coated with silver, deposited on the surface from a chemical solution. The process was quite simple provided the worker was meticulous in his preparations and was willing and able to follow directions exactly. In essence, it involved making up a solution of silver nitrate, changing the silver nitrate into silver oxide by the addition of ammonia water and potassium hydroxide, and then removing the oxygen from the silver oxide by the use of a reducing agent such as dextrose. The mirror was placed face up in the solution, and when the reducing agent was added, the liberated silver was deposited on the face of the mirror. This method of silvering mirrors is still used for producing test surfaces on various types of mirrors.

Unfortunately, a silvered surface is far from permanent since it is readily attacked by many impurities ordinarily contained in the air. Perhaps the most damaging of these impurities are the sulfides (sulfur compounds), which produce the same tarnish on the mirror's surface as that which appears on table silverware. Because the layer of silver on a mirror is so thin, the tarnish cannot be removed by ordinary methods without permanently damaging the coating. Consequently, a silvered mirror will give good service for only about six months, after which the old coat must be removed and replaced with a new one.

The reader may wonder about this impermanence when he considers the mirrors in his home, some of which may have belonged to his great-grandfather and which are still as bright and reflective as when they were made. The difference is that ordinary mirrors are silvered on the back. The silver is then covered with a coat of paint, which effectively seals it off from the damaging effects of the atmosphere. Astronomical mirrors, however, are silvered on the front, or curved, surface (they are called *first-surface* mirrors because of this).

Because of the difficulty of caring for a silvered surface, astronomical mirrors and diagonals (the flat secondary mirrors that we consider in the next chapter) are no longer coated with silver. Instead, mirror makers have turned to another metal, aluminum. Aluminum does not *appear* to be as reflective as silver or some of the other metals, such as chromium and nickel, but in fact the difference is very slight. The light to which the eye is most sensitive has a wave length of about .59 microns, and the table below shows how the various metals compare at this wave length.

METAL	LIGHT REFLECTED (per cent)
Aluminum	83
Chromium	70
Iridium	75
Nickel	62
Platinum	59
Rhodium	78
Silver	93
Speculum metal	64

Almost any of the above metals might make satisfactory mirror coatings from the point of view of reflectivity, but aluminum has an additional quality (besides its cheapness) which makes it particularly adaptable to this purpose. It is quite readily attacked by oxygen, and the substance produced, aluminum oxide, forms a protective coating over the aluminum underneath. Furthermore, pure aluminum oxide is colorless—you have seen it in the form of rubies and other precious stones, but here the color is produced by various impurities—and is also very hard. Consequently an aluminum coating on your mirror, properly cared for, will last for six years instead of six months, as in the case of silver. Some mirror makers attempt to protect the reflective coating further by covering it with a thin layer of silicon monoxide, quartz, or one of the fluorides. The disadvantage here is that such overcoatings are very difficult to remove without damaging the mirror surface below. A compromise is to use one of the aluminum alloys—the so-called Beral coatings, made up of aluminum and beryllium, are probably the best—since these alloys are tougher and longer lasting than aluminum itself.

Unfortunately, aluminum is much too active chemically to be produced from solution by ordinary chemical means. It must be vaporized by heating it in a high vacuum. The vapor deposits on the mirror to form a uniform, thin coating. The means of producing a vacuum good enough for this purpose are too complicated and costly for most amateurs, so this is one process in the making of a telescope which is best done by professionals who have the necessary equipment. The result is well worth the cost, which is about $10–$20 for an 8-inch mirror. (See appendix for a list of aluminizing services; also how to pack your mirror for shipment.)

You can test the coating for the proper thickness by holding it up to a 100-watt bulb. The light should show through faintly, with a few minute bubbles visible in the aluminum coat. If the coating is thicker than this it tends to spoil the figure of a good mirror. If it is thinner, some of the light striking its surface will be transmitted rather than reflected.

CHAPTER 10

The Diagonal

Much of the hard work that you have put in on your mirror will be wasted unless you make or buy a diagonal whose optical qualities are of the same excellence as the mirror itself. The function of the diagonal is to pick up the reflected rays from the mirror and shunt them at a right angle into the adapter tube of the eyepiece. In order to accomplish this objective, you must have a surface of the diagonal as flat as you can make it—within $\frac{1}{8}$ wave of light, if possible. The actual production of such a surface is not very difficult physically, but the preparations to do the job are lengthy. For this reason many amateurs prefer to buy their diagonals—they cost about $20 for an 8-inch mirror of medium focal length—and devote their energies to the remaining parts of the telescope mounting. But whether purchased or homemade, the diagonal must fulfill certain specific requirements of size, shape, and positioning, so let us consider these requirements first.

PLACING THE DIAGONAL

As can be seen from the diagram, the diagonal is simply a turning point for the rays coming from the surface of the mirror. It has no effect on the focal length of the mirror since it causes the rays neither to diverge nor to converge from their prescribed path. The distance d from the center of the diagonal to the point where the reflected rays would normally come together to form the focal plane is exactly equal to the distance d' to the new focal plane with the diagonal in place. The only

131

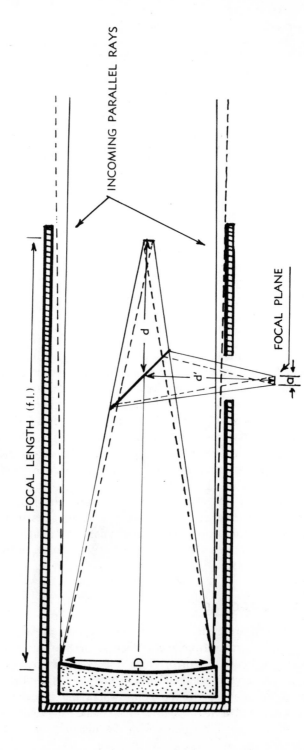

Placing the diagonal.

problem, then, is to make d' long enough to reach and pass through the side of the tube, plus 3 inches to permit adequate clearance on the outside of the tube for eyepiece or camera. Many amateurs wish to record on film what they see through the eyepiece; this requires a little more space than if the telescope is to be used solely for visual work since clearance must be left for the rays to pass through the camera body and focus on the film.

Therefore the limiting factors as to where the diagonal must be placed are the focal length of the mirror, the diameter of the telescope tube, and the extra 3 inches for clearance. Since the mirror will have a focal length of 56 inches and will be enclosed in a tube whose outside diameter is 9.5 inches (for reasons we explain later), d' will be 7¾ inches and the diagonal will be placed 48¼ inches from the mirror. Strangely enough, the diagonal should not be exactly centered in the tube, but should be offset a little in a direction toward the mirror end of the tube. A careful look at the geometry of the diagram will show you why this is true. But the amount of offset is not enough to influence the dimensions indicated above, and can be compensated for at the eyepiece itself.

TYPES OF DIAGONALS

The most easily maintained diagonal is a right-angled prism with a 45° face because all surfaces are glass and are therefore easy to clean. The light enters one of the 90° faces (see diagram), is totally reflected from the inside surface of the hypotenuse face, and leaves the prism through the other face. The loss of light is only 10 per cent for uncoated prisms; and, if the surfaces are coated with magnesium fluoride or other antireflective materials, this loss may be cut to as low as 1 per cent.

Yet there are disadvantages to the use of a prism. The surfaces must necessarily be square, which cuts off an appreciable portion of the light entering the tube. The faces must be oriented exactly 90° to one another, the glass must be of the highest optical quality and must be completely homogeneous, and all three surfaces must be plane to within ⅛ wave. Few prisms satisfy these requirements and those that do are likely to be expensive. Finally, it is much more difficult to mount a prism than a flat diagonal mirror. In spite of these qualifications, many amateurs still use prisms in their telescopes, partly from ignorance but chiefly because they are easy to obtain and, once installed, require practically no upkeep and a minimum of attention. If you decide to join this group, be sure you specify the use the prism is to be put when ordering. Many supply houses have large supplies of tank prisms (used for periscopes in Army tanks) which are cheap but completely useless for optical purposes. There are, sadly enough, quite a few amateurs who have ruined the performance of excellent mirrors by trying to use these things.

Prism mounted on a 12-inch open-tube telescope. The ring holding the prism and eyepiece can be turned to make viewing more comfortable. Because of its exposed position, a prism is better adapted to this type of mounting than is a diagonal mirror.

Diagonal Mirrors

A better alternative to a prism is a flat diagonal mirror, aluminized on its surface. The diagonal is inclined 45° to the axis of the mirror, and is located as outlined earlier.

Your primary concern, whether you use a prism or a diagonal, is how large it must be. You will want to catch as many as possible of the reflected rays from the mirror, but if you make the diagonal too large it will cut off a disproportionate amount of the light entering the telescope. It will also produce unpleasant diffraction effects.

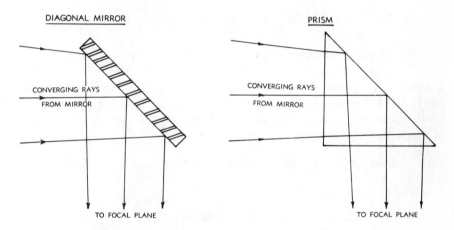

Although the prism and the diagonal serve the same purpose in the optical train, the prism reflects light from an inside surface (the hypotenuse), while the diagonal reflects from an aluminized outside surface.

DIAGONAL MATHEMATICS

Here we must digress a little to talk about eyepieces, since they have great influence on the size of the diagonal. Eyepieces usually have two main lenses: the field lens, which picks up the image formed by the telescope mirror at the focal plane, and the eye lens which alters this image so that it can be seen by the eye. Other things being equal, low-power eyepieces produce small images but a wide field of view, while high-power eyepieces have the opposite effect. The diagonal must be large enough to reflect a cone of light the width of the field lens of your lowest-power eyepiece, since this will determine the maximum field of view of your telescope. The diameter of the field lens of such an eyepiece is about $7/8$ inch.

The width of the diagonal, based on this measurement, can be found from the formula

(1)
$$w = \frac{d(D - a)}{f} + a$$

where D = diameter of mirror
  f = focal length of mirror
  a = width of focal plane (or width of field lens of eyepiece)
  d = distance from focal plane to diagonal

Knowing these measurements, we can determine the width (w) of the diagonal for an 8-inch mirror of 56-inch focal length as follows:

$$w = \frac{7\frac{3}{4} \, (8 - \frac{7}{8})}{56} + \frac{7}{8} = 1\frac{7}{8} \text{ inches}$$

Many amateurs prefer to approach the problem from another angle, however. Since the moon is the most commonly observed object in the heavens, they prefer to place the diagonal in such a position that the image of the moon just covers the focal plane. The angular diameter of the moon is 31 minutes of arc, which may be transformed into inches at the focal plane by another formula:

(2)   diameter of focal plane = tan 31′ × focal length of mirror;

but tan 31′ is .009, so this becomes

$$a = .009 \times 56 = .504 = \frac{1}{2} \text{ inch}$$

Substituting this value for the diameter of the focal plane in equation (1) above, we get

$$w = \frac{7\frac{3}{4} \, (8 - \frac{1}{2})}{56} + \frac{1}{2} = 1.57 = 1\frac{5}{8} \text{ inches}$$

This is a more workable dimension from the point of view of cutting off light from the mirror. Assuming a diagonal of this width, let us see how much light is actually lost. The width of the diagonal, 1.57 inches, must be multiplied by 1.4 to obtain its length, since the glass will be tipped at 45°. If the diagonal is made elliptical, it will cast a shadow whose area will be $\pi(1.57/2)^2 = 1.92$ square inches on the center of the mirror.[1] The area of the whole mirror is 16π, or 50 square inches. Thus the diagonal cuts off 3.8 per cent of the light. But if the diagonal

[1] For those who wonder how an elliptical mirror can cast a round shadow, remember that an ellipse tipped 45° will act as a circular obstruction to the passage of light. Prove this for yourself by cutting from paper an ellipse whose long axis is 1.4 times its short axis. Tip the long axis at 45° to the line of sight and squint at it with the eyes nearly closed. It will look like a circle.

is rectangular instead of elliptical, its area is even greater: 2.46 square inches, which will cut off 4.94 per cent of the light.

In general, it is better to make a diagonal too small than too large, at least for visual observations, for two reasons. As pointed out above, it is desirable to cut off as little light as possible, and also because a diagonal whose shadow area is greater than 6 per cent of the total mirror surface introduces diffraction effects in the images. These diffraction effects are also a product of sharp corners or edges on the diagonal, which is why it is best to make it elliptical in shape. But if you plan to use your telescope for photographic purposes, the 1⅞-inch diagonal mentioned above will be more useful than the smaller size. This is because you will probably wish to cover as large a field as possible in taking photographs of the heavens. Most of the large nebulae, for example, cover a field of more than 31 minutes. Some amateurs make *two* diagonals, one for visual work and one for photography.

TESTING GLASS FOR FLATNESS

As mentioned earlier, the surface of the diagonal must be as flat as you can make it, preferably within about ⅛ wave. The fact that the light is reflected from the diagonal at an angle magnifies the effect of any distortions on its surface. Lest this insistence on flatness discourage you from making your own diagonal, let us hasten to add that it is easier to make and test a diagonal mirror of flat surface than it was to make the primary curved mirror.

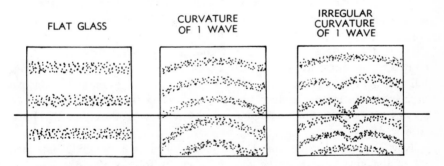

FLAT GLASS	CURVATURE OF 1 WAVE	IRREGULAR CURVATURE OF 1 WAVE

Testing glass for flatness, using interference fringes. Each of the three surfaces shown above is being tested against a reference blank assumed to be flat.

In a common high school experiment in physics you take two supposedly flat pieces of glass, hold them under a bright light, and squeeze them together. If the pieces are flat enough faint colored rings can be seen on the inner surfaces. If the source of light is the sun, or an ordinary tungsten light bulb, the rings will be shades of red and green and will be poorly defined. But if a sodium flame is used (produced by adding salt to the flame of a Bunsen burner), the rings will be alternately yellow and black and will appear much more prominently. These are called Newton's Rings, so called because Sir Isaac Newton first observed them. They appear because the light from the lamp or Bunsen flame is bent slightly as it passes through the thin layer of air between the glass plates. The thickness of this layer of air is, of course, a measure of the flatness of the glass surfaces as they are pressed together. Its thickness also determines how much the light is bent. The lines (called fringes) appear when the light is bent just enough to make the waves interfere with each other.

Now if you use two pieces of very flat glass, the rings are so large that you can no longer see complete circles, but only arcs. The "bull's-eye," or center of the circle, disappears. Finally, with perfectly flat glass, the fringes become completely straight. The straightness of the fringes can be used as a direct measure of the flatness of the surfaces. If you lay a straight edge parallel to the general direction of the fringes, the number of lines it intersects indicates concavity or convexity on the glass equal to half that number of wave lengths of light. Thus if the straight-edge cuts across 2 fringes, there are curves on the glass equal to 1 wave length high or deep. If the straightedge cuts no lines, but simply approaches another, the relative amount of approach may be used as a measure. For example, if the straightedge crosses approximately half the distance between 2 fringes, curvature equal to $\frac{1}{4}$ wave length is present.

Further experiment with the glasses makes another characteristic evident. If the two glasses are concave toward each other, pressure at the edges causes the fringes to move toward the center (as deduced from the way the fringes curve) and become greater in number. If they are convex, edge pressure produces the opposite effect. In short, we can tell very quickly whether the surfaces of two pieces of glass are concave or convex toward each other simply by pressing them together and observing the direction and number of the fringes produced as the pressure is applied.

Now if you know that one of the pieces of glass is perfectly flat, then all of the curvature must be in the other. It follows that the simplest way to test a diagonal for flatness is to check it against a reference flat. Unfortunately, reference flats are more expensive than the diagonals they test. Hence it would be better to buy a diagonal of specified flatness in the first place, unless, of course the reference flat is purchased by a club for the purpose of testing many diagonals.

For the individual worker, however, there is a way out of the difficulty. Suppose you obtain three pieces of glass which you will mark *X, Y,* and *Z* with a china-marking pencil. Test each against the other two, recording your results. For the sake of convenience, indicate the number of fringes of concavity as a negative number, and convexity as positive. You will end up with three equations, as the following example illustrates.

(1)	$X + Y = 2$ (convex)
(2)	$X + Z = -2$ (concave)
(3)	$Y + Z = -3$ (concave)

Solving simultaneously,

(1)	$X + Y = 2$
(2)	$X + Z = -2$
(4)	$\overline{Y - Z = 4}$ [subtracting (2) from (1)]

Now you can use this result with equation (3) above, or,

(4)	$Y - Z = 4$
(3)	$Y + Z = -3$
	$\overline{2Y = 1}$ [adding (4) and (3)]
or	$Y = \tfrac{1}{2}$
Then	$X + \tfrac{1}{2} = 2$ [substituting in (1)]
or	$X = 1\tfrac{1}{2}$
Finally,	$\tfrac{1}{2} + Z = -3$ [substituting in (3)]
or	$Z = -3\tfrac{1}{2}$

The outcome of all this is that the X glass is $\tfrac{3}{4}$ wave convex, the Y glass is $\tfrac{1}{4}$ wave convex, and the Z glass is $-\tfrac{7}{4}$ wave concave, assuming there are no serious local irregularities on their surfaces. Since the Y glass is most nearly plane, you will keep this one for a reference surface and attempt to grind the X glass to flatness.

The principle here is relatively simple; putting it into practice is a little more difficult. For best results, a monochromatic light source must be used because this yields the clearest and sharpest fringes. Furthermore, a light of a single color gives precise information because only a single wave length is used, whereas a tungsten lamp or ordinary sunshine covers the whole range of the visible spectrum and you can't be sure which wave length the fringes represent. The simplest arrangement is to use a neon lamp in a darkened room. Neon light bulbs can be purchased in almost any hardware store. Buy the bulb type rather than tube arrangement since the former can be screwed into any standard lamp socket and is also a more convenient size.

Set up an arrangement such as shown in the diagram. The frosted glass at the top serves to diffuse the light from the neon bulb, thus providing more even illumination. The $\tfrac{1}{4}$-inch polished plate glass in the middle is placed at a 45° angle and acts as a "beam-splitter." Most of the light from the neon bulb passes through, but some of the reflected light from

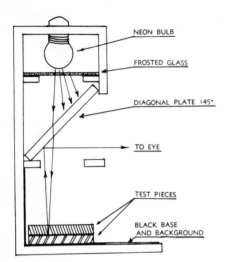

NEON BULB

FROSTED GLASS

DIAGONAL PLATE (45°

TO EYE

TEST PIECES

BLACK BASE
AND BACKGROUND

Schematic drawing of a possible set-up for testing for flatness.

the test surfaces is reflected again from the lower surface of the 45° glass and enters the eye. The pieces of glass to be tested are placed on a piece of black cloth or cardboard. The advantage of this device is that both light and eye are almost perpendicular to the surfaces to be tested. This is important because if the fringes are viewed obliquely, the refraction of the glass will cause serious errors. Make a straightedge by blackening a plastic ruler.

Plate glass is the best source of nearly flat surfaces for diagonals. It usually comes in ¼-inch thickness, but if you are lucky you may be able to pick up a piece ⁵⁄₁₆ or ⅜ inch thick. Glass thicker than this is usually not well annealed, that is, it has not been heat treated sufficiently to relieve all the strains in it and may shatter or warp when cut. On the other hand, thin glass has a tendency to bend out of shape in the grinding and polishing processes, during which it is cemented to a supporting plate. When the cement is removed, the glass may return to its original form, leaving the surface no better than it was in the beginning. Wrecking companies and junk dealers are fruitful sources of good plate glass; we once found some nearly ½ inch thick, perfectly annealed and beautifully flat, which had been used in a shower stall in an old house. Other sources of flat glass are war-surplus optics, old camera filters, and the like.

CUTTING THE GLASS

Cutting thick plate glass is a problem. The best solution is to take it to an experienced glazier and have him cut it into the sizes you wish. Lacking that, you can do it yourself if you are very careful and thorough in your preparations. Borrow or buy a heavy-duty glass cutter, one which produces the deepest possible groove in the surface of the glass. The kind

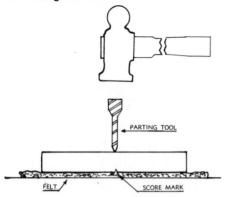

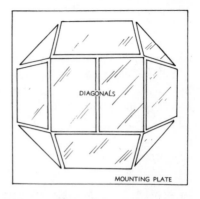

Cutting the diagonal sections.

Mounting the sections for grinding and polishing.

with a carbide wheel as the cutter is the best. Lay the glass on a flat surface and clamp a straightedge along the line you wish to cut. It is wise to mark the glass into the projected pattern of cuts with a china pencil first. Score the glass heavily with a *single* stroke of the glass cutter, since running the cutter back and forth along the groove produces a line of uneven depth and is conducive to fracturing the whole piece of glass. Do one cut at a time. After you score the glass, turn it over and place it on several thicknesses of flannel or other soft cloth. A piece of Celotex or other soft fiberboard makes a good base if flannel is not available.

A parting tool—an implement which looks like a wide chisel but has a rounded, rather than a sharp, edge—is now used. Place the edge directly over the score mark as seen through the glass and strike the end of the tool a sharp blow with a light hammer. If you are fortunate, the glass will part cleanly and evenly, but you may have to clean off chipped edges or sharp projections before making the next cut. If you find it impossible to obtain a parting tool, you can make one yourself by rounding off the edge of an old chisel, if you can find one more than 3 inches wide. Or you can round off the edge of a piece of flat steel, $\frac{1}{16}$ inch thick and 4 inches long. Place the rounded side on the glass and strike the other side with the hammer. In any event, what you are trying to do is to produce a sudden pressure along the scored line on the glass in the hope that the supporting flannel or Celotex will give the remainder of the glass enough support to cause it to break along the line. Try out the whole procedure on a piece of thick window glass first.

Cut out three pieces each 2×3 inches. Mark the back of each piece with a china pencil. Then cut out enough additional pieces to place around two of these as indicated in the diagram. These additional pieces will help to support the grinding tool so that the edges of your diagonals will not be turned during the grinding. Wash all pieces in a strong detergent after beveling their edges slightly on a piece of scrap glass, using No. 120 Carborundum and water as an abrasive.

Test the three pieces in pairs under the neon lamp, pressing them together until the interference fringes spread out to a distance of about $\frac{1}{4}$ inch. In carrying out your tests, be sure that each pair is checked under the same conditions, that is, that they occupy the same position and that the eye is placed in the same spot for each individual test. Be sure also that temperature conditions are the same for each test: do not handle them any more than is strictly necessary, since the warmth of the fingers has a very considerable effect.

GRINDING AND POLISHING THE DIAGONAL FLAT

For this operation you will need two pieces of plate glass, one 6×6 inches and the other 8×8 inches. You should also cut out a circle of $\frac{1}{2}$-inch plywood, 8 inches in diameter and seal its surfaces with shellac or plastic spray.

Warm the 6×6 piece of glass in a bucket of water to about 150° F. Remove it from the water, dry it with an old towel, and paint one surface with a thin layer of melted paraffin. Set the two diagonals in the center of the plate and surround them with pieces cut to form a rough circle, as shown in the diagram. Press each piece down firmly, leaving $\frac{1}{16}$-inch spaces between them. Fill in these spaces with melted paraffin. Scrub the exposed surfaces with a detergent.

Lay the 8×8 piece of glass on a flat surface. Sprinkle it with No. 400 abrasive, wet it down with the spray bottle, and grind the mounted pieces of glass against it. Use 2-inch "W" strokes, taking care that the pieces of glass on the outside do not protrude over the edge at the end of the stroke. Grind for six wets, reversing the grinding plate and diagonal block with every wet. All pieces of glass should be "frosted" after six wets, or at least after another six wets. If not, you have mounted them unevenly on the surface of the plate glass. It will be better in this case to remove and remount them.

After they are all frosted, wash thoroughly under a stream of water and examine the spaces between the mounted blocks to see that all grains of No. 400 have been removed. If necessary, scrape out these spaces with the point of a knife and reseal with more melted paraffin.

Repeat the whole process with No. 600, then with No. 900, and finally with 305E, taking care between each grade that none of the previous grade is left in the spaces between blocks. A few extra wets with 305E will do no harm.

Wrap a strip of waxed paper around the rim of the shellacked disk of plywood, secure its ends by passing a stout rubber band all the way around its circumference, and pour on a lap of hard pitch. The pitch should be between $\frac{1}{8}$ and $\frac{1}{4}$ inch deep for this lap. When it cools sufficiently, strip off the paper collar and press with the ground surface of the 8×8 glass plate, using a rouge mixture to keep the plate from sticking to the lap. For this pressing place a piece of flat board over the glass and distribute weights (10 to 20 pounds) evenly over the board. Leave the weights on until you are sure that glass and lap are making good contact. Since the spaces between the glass blocks provide adequate channels for air and polishing compound, this lap need not be faceted, although it is a good idea to cut one or more deep channels all the way across the face of the disk. Lay on a section of plastic screening and cold press until the pattern of the screening is well established in the pitch.

Charge the lap with a thick mixture of Barnesite or cerium oxide and polish with the diagonal blocks on top. Again use 2-inch "W" strokes and avoid overhang. After 15 or 20 minutes of polishing, wash up everything. Wait for the diagonal blocks to cool, and test both diagonals with the reference block while they are still in place on the mounting plate. They should test within a single wave length; if not, you must return to fine grinding.

This test is carried out just as the previous tests were except that you now know the curvature of your reference blank. If the reference blank is, say, ¼ wave concave, straight fringes would indicate a convex diagonal. But a flat diagonal will *appear* concave when tested against the concave surface of your test blank. If the fringes indicate ¼ wave concavity, the diagonal you are testing must be flat and the deviation in the fringes must all come from the test blank. *More* than ¼ wave concavity would indicate that the diagonal is concave by an amount equal to the difference between the test figures and the known amount on the test blank. Check the distances between fringes and the partial distances between fringe and straightedge with dividers. All tests must be done carefully and cautiously for the polished surfaces scratch very easily.

Some authorities recommend reversing tool and diagonal block with every wet to prevent the formation of concave surfaces on the diagonals. This is rarely necessary if the strokes are kept short.

Continue polishing for half-hour spells until the surfaces are completely polished. Allow the diagonal block to cool for half an hour before testing. When the surface on each diagonal tests to ⅛ wave or better, remove them from the block by immersion in water and slow heating. Be very careful here; one drop of cold water on hot glass surfaces can shatter your beautifully polished and figured diagonals. Remove any remaining paraffin with turpentine, wash in warm water with a good detergent, and allow the diagonals to cool for an hour before making a final test. Test the two diagonals against each other, again using great caution because these surfaces scratch unbelievably easily.

There is only one possibility of misfortune at this point. If the glass in the diagonals has been warped by the grinding and polishing action while cemented to another surface, there is a slight possibility that their surfaces will become astigmatic when strains are relieved by removing the cement. This is the advantage of making two diagonals at a time, since it is unlikely that both will be affected this way. If both turn out well, save one for a reference flat for future use. Wrap it in absorbent cotton and store it with the tool and pieces of plate glass in a dust-free box or cupboard. The chances are that you will want to use the apparatus in some future operation.

SHAPING THE DIAGONAL

There are several ways of shaping the rough edges of the rectangular diagonal into an oblique elliptical curve. The one that requires the least labor and is also safest is the use of a drill press and what mirror makers call a "biscuit cutter."

Sandwich the diagonal between two pieces of plate glass, using a 1-3 mixture of beeswax and rosin as a cement. Make a small box of plywood and fit the sandwich into the box at a 45° angle, using cleats to hold it firmly in position. Mark the final position of the mirror carefully on the outside of the box. Now fill all the spaces around the sandwich with plaster of Paris, allowing at least 24 hours for the plaster to dry.

The biscuit cutter can be made from a 4-inch section of thin-walled brass tubing whose inside diameter is $\frac{1}{16}$ inch larger than the desired small diameter of the ellipse ($1\frac{7}{8}$ inches). Cut a cylindrical plug 1 inch long from well-dried oak or maple of a diameter to insure a very snug fit inside the tubing. Tap this into one end of the tube. Punch-prick the center of the wooden plug (the center of a circle may be found by drawing two diameters at right angles to each other), and drill a $\frac{1}{4}$-inch hole. Insert a $\frac{1}{4}$-inch shaft made by cutting off the head of a bolt, and secure it in place by drilling a hole through tubing, wood plug, and shaft, and inserting a pin. Using a triangular file, cut deep serrations in the open end of the tubing. Then drill several $\frac{1}{8}$-inch holes in the sides just below the wooden plug to provide ventilation. Install the shaft in the chuck of the drill press, and true the whole arrangement until there is no wobble at slow speeds (100 to 200 rpm).

Clamp the box containing the diagonal to the drill-press platform. Check your measurements to be sure that the biscuit cutter will pierce the diagonal at the proper place. Now place aluminum foil, Saran wrap, or some other protective covering over all parts of the drill press, since the abrasive used in this operation is ruinous to machine tools if it gets into any working parts.

Using a mixture of kerosene and water with No. 120 abrasive applied to the business end of the cutter, start drilling. Don't hurry this job; the action of the abrasive and metal on the glass after the cutter has pierced the plaster of Paris will generate lots of heat. Lift the drill often

A "biscuit-cutter" being used to perforate a Cassegrainian primary mirror. The protective cover was removed from the drill press while the picture was being taken.

to permit cooling. Add abrasive often, too, since it is broken down rapidly by the stringent action in the narrow confines in which it is operating. Excessive heat may break the seal in the glass sandwich, and of course if a single grain of abrasive works its way into the space between protective plate and diagonal, your carefully nurtured surface will be ruined. Yet it is sometimes astonishing how much abuse glass will stand. One of our students once drilled all the way through a mirror, stopping only to add more abrasive and kerosene. He finished the job in half the time ordinarily required. In spite of the fact that the serrations were worn down to the hub and the whole arrangement was smoking hot, the hole was smooth and clean as a whistle.

After the drilling job is completed, remove the glass sandwich from the plaster matrix carefully. Soak it in turpentine to remove the protective plates from the diagonal. Clean everything up, and grind a narrow bevel all around the surface edge of the diagonal if it became chipped in the cutting process. This bevel will have to be painted flat black after the diagonal comes back from the aluminizers. If the edge is still good, beveling is unnecessary. Test the finished product once more against its mate to be sure the drilling hasn't relieved any strains in the glass which might affect the surface. Put it aside to be sent off with the main telescope mirror for aluminizing.

CHAPTER 11

The Eyepiece

An EYEPIECE is a combination of lenses whose primary function is to enlarge the image formed by the objective lens of the refracting telescope or the mirror of the reflector. A single lens element could act as an eyepiece, but it has so many deficiencies that it is seldom used except in the cheapest telescopes. In order to obtain the quality we require in a good eyepiece, we may use as many as six or seven lens elements. These may be individually mounted or cemented together, depending on the characteristics desired for the eyepiece. Ordinarily, there are two main lenses: the *field lens,* which picks up rays from the mirror or objective, and the *eye lens,* which directs them to the eye. Many of the important characteristics of the telescope depend on the particular eyepiece used, so this part of the instrument deserves some careful attention. Let's start by listing some of the terms used to describe eyepieces.

FOCAL LENGTH

The focal length of a simple lens may be found by finding the distance from the lens at which the rays from a distant object will produce a sharp image. Later in the chapter we describe the method used for an exact determination of focal length. The focal length of a combination of lenses, such as is used in an eyepiece, is more difficult to find. In the case of the so-called negative eyepiece, where the image is produced within the eyepiece itself, it is impossible to find the focal length by experiment and we must depend on a formula which makes use of the focal lengths of the individual lens elements. In order to distinguish between the focal length of the eyepiece as a whole and that of any of its elements, we use the term *equivalent focal length* for the eyepiece itself, meaning that this combination of lenses acts as though it were made up of a single lens whose focal length we know. This formula is

$$efl = \frac{f_1 \times f_2}{f_1 + f_2 - d}$$

where *efl* = equivalent focal length
  f_1 = focal length of field lens
  f_2 = focal length of eye lens
  d = separation of lenses

It is always necessary to know the focal length of the eyepiece since the magnification (or power) of the whole telescope depends upon it.

MAGNIFICATION

Far too much importance is attributed to this characteristic of the telescope. Most beginners believe it is the essential factor in the performance of a telescope, that enlarging the image is the most important aspect of good seeing. Actually, there are many cases where magnification is undesirable. On nights of poor seeing, when the turbulent atmosphere causes images to shake or jump around in the field, high magnification is useless. In any event, there is no point in using powers greater than those needed to exhaust the resolving power of the telescope except in those rare cases when we might wish to separate the components of a double star. As we saw previously, resolving power is the ability to pick out detail in an image. To do this, we often need increased magnification. But if we use more than we need to separate detail, we injure the clarity of the image rather than enhance it. The photographer who has a fine-grain, sharp negative finds that an enlargement improves the quality of the picture he produces; but he also knows that, beyond a certain point, "blowing up" his negative does more harm than good.

Magnification, to use terms which are a little more technical, consists of increasing the angular diameter of the image. A distant tree, for example, has a small angular diameter. As we walk toward it the angle between top and bottom increases, which is what makes the tree seem to increase in height. In the same way, the mirror of the telescope produces an image whose small angular diameter may be increased by the eyepiece. The amount of increase depends on the relationship of the focal lengths of mirror and eyepiece and is equal to the first divided by the second. Thus a mirror whose focal length is 56 inches will produce a magnification of 56 times if used with an eyepiece of 1-inch *efl* (25.4 mm), but the same mirror will magnify an object 224 times if the eyepiece used has a focal length of ¼ inch (6.4 mm).

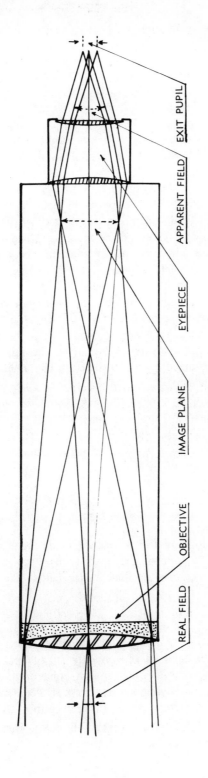

REAL FIELD   OBJECTIVE   IMAGE PLANE   EYEPIECE   APPARENT FIELD   EXIT PUPIL

Elements of a telescope.

Although this diagram shows a refractor, the terms used and the way they are measured are common to all telescopes.

REAL FIELD

The field of view of a telescope is the angular diameter of the area which the telescope can survey. Sometimes, for terrestrial telescopes, this is given in terms of linear width at a given distance, for example, 469 feet at 1000 yards. Designations such as this are usually applied to binoculars, spotting scopes, and field glasses, in which the eyepiece is a fixed part of the instrument. But in astronomical telescopes, which use a variety of eyepieces to change the magnification, the field size is classified by angular measurement.

The angle formed by converging rays brought to a focus by the mirror of a telescope represents the potential field of the instrument. The part of the potential field which can be utilized by the eyepiece is called the *real field,* or *field of view,* of the telescope. The intersection of rays from all parts of the mirror forms an image at the *focal plane,* and the angular size of the real field is determined by the part of the focal plane which can be picked up by the eyepiece.

The eyepiece also produces a cone of convergent rays, and the angle at the intersection of these rays is called the *apparent field.* But the rays picked up at the eyepiece, of course, come from the mirror. The result is that the angle of the converging rays from the mirror is changed into another, and wider, angle by the eyepiece. The mathematical ratio between the two angles gives us another way of expressing magnification, for the eyepiece simply spreads out the image by an amount equal to this ratio. Thus we can write:

$$\frac{\text{apparent field}}{\text{real field}} = \text{magnification}$$

or

$$\text{real field} = \frac{\text{apparent field}}{\text{magnification}}$$

If what we are looking for is size of field, then, we must keep the magnification low. It follows that long-focus eyepieces which produce low magnifications will also produce wider fields of view.

EXIT PUPIL

The size of the cylinder of light rays which enters a telescope depends, of course, on the diameter of the objective. The optical elements in the telescope bend these rays together into a cone until, when they emerge from the eyepiece, the cylinder has become very much smaller. This emergent cylinder of rays is called the exit pupil, and its cross section is known as the Ramsden disk. The amount the rays are bent together or condensed depends on the focal lengths of mirror and eyepiece, and this relationship has already been expressed as magnification. Thus we may express magnification in another way:

$$\text{magnification} = \frac{\text{diameter of mirror}}{\text{diameter of exit pupil}}$$

The diameter of the exit pupil is relatively easy to measure. If the telescope is directed at the sky or at any bright object such as a sunlit white wall, the Ramsden disk may be clearly seen as an image of the objective. It appears to float in the air just behind the eye lens of the eyepiece. You can find the diameter of the Ramsden disk by placing a piece of ground glass behind the eyepiece and then measuring the circle of sharpest focus with a pair of calipers. This is very useful in finding the equivalent length of an eyepiece where the focal lengths of the component lenses are unknown. Since the magnification of the telescope as a whole equals objective divided by exit pupil, and since the focal length of the eyepiece equals the focal length of the mirror divided by magnification, the relationship can be easily worked out, as in the following example.

diameter of mirror = 8 inches
diameter of exit pupil = $\frac{1}{7}$ inches (by measurement)
∴ 8 ÷ $\frac{1}{7}$ = 56 = magnification
and if the focal length of the mirror is also 56 inches, then
56 inches ÷ 56 = 1 inch = focal length of eyepiece

But the exit pupil is very important for another reason, which has to do with the eye itself. The pupil of the eye in daylight varies from 2 mm to 4 mm, depending on the brightness of the day. At night, however, it expands to 7 or 8 mm. An exit pupil of 7 mm, therefore, cannot possibly enter the eye during daylight, and the consequence is a waste of that part of the emergent beam of light which the eye pupil cannot accommodate. Because the exit pupil consists of light from the whole mirror, cutting off the outside portion of the exit pupil is equivalent to reducing the diameter of the mirror. Thus if the exit pupil is 8 mm and the eye pupil is 4 mm, only half the diameter of the mirror is in use. On the other hand, if the exit pupil is only 1 or 2 mm, much of the potential of the eye itself is wasted during nighttime observation. High-powered eyepieces of short focal length have small exit pupils, which is another way of saying that high magnifications cause loss of light and a consequent diminution of the brightness of image. A rough rule for the estimation of comparative light-gathering power of an eyepiece is to use the square of the diameter of the exit pupil. In the same telescope, two eyepieces having exit pupils of 2 and 4 mm, respectively, will have light-gathering powers of 4 and 16 in relation to one another, or a ratio of 1 to 4.

EYE RELIEF

The distance from the eye lens to the point where the eye can best observe the image produced is called eye relief. This distance is usually a little less than the effective focal length of the eyepiece and can be found from the formula

$$ER = bfl + f_e^2/f_o$$

where bfl = back focal length, which is the distance from the last surface
         of a lens or a system of lenses to the point where incoming
         parallel rays have been bent to a focus
    f_e = focal length of eyepiece
    f_o = focal length of objective

Poor (short) eye relief and small exit pupil go together since each is the result of high magnification. This combination is likely to be tiring for the observer since he must hold his eye close to the eyepiece and, because of the small exit pupil, must hold his head very steady to keep the image centered in his vision. It is especially bothersome to those who wear glasses. Usually, in this case, it is better to remove the glasses and compensate for any deficiency in the focusing of the eyepiece.

From a consideration of the factors mentioned above it should be evident that high magnifications have as many disadvantages as virtues. For comfortable observing under ordinary circumstances, low powers are best. With a 56-inch focal length of the mirror, a focal length of ½ inch (12.7 mm) for the eyepiece will give you a magnification of 112, which is comfortable and easy to use and brings out most of the detail you wish. You will probably be surprised to find that a magnification of only half the above, using a 1-inch (25.4 mm) eyepiece, will produce some of the most enjoyable sights in the heavens because of the wider field which accompanies it. The new moon, open star clusters, eclipses, and large nebulae are more beautiful objects when seen in their entirety than when you can see only part of them.

A rule of thumb states that 50 magnifications per inch of aperture is a limit beyond which images become hopelessly dim and indistinct. This may be true for the planets and other expanded bodies, but it does not necessarily apply to stars, particularly double stars. On nights of good seeing when there is little or no turbulence, much higher powers can be used to advantage. In the long run it depends upon the judgment of the observer.

GENERAL NOTES ON EYEPIECES

There are many different types of eyepieces, each characterized by a particular choice and arrangement of lens elements, and each with its own advantages or deficiencies. In fact, there are at least 67 different combinations of lenses which may be used for eyepieces.[1]

Complexity in an eyepiece is not synonymous with excellence of performance. Some relatively simple ones work as well as their complex and expensive cousins, depending on the individual telescope and the purpose for which it is used. Allowing for this, the real criterion for eyepiece quality is the care with which they are manufactured. For this reason, it is best to buy them from manufacturers whose product has stood the test of time. This is not to say that the many eyepieces offered by dealers from war-surplus stocks are no good, for quite often the contrary is true and many of them are excellent. Unfortunately, it is rarely possible to test an eyepiece before buying it, and you may find to your sorrow that the beautiful eyepiece which cost the government so much is good for little except a general view of the moon, where almost any eyepiece will do. Generally speaking, government surplus eyepieces were designed for special optical systems and, when taken out of those systems, do not perform to their best advantage. Most of the surplus eyepieces perform best in very short focal length systems.

In general, eyepieces are subject to a half dozen defects, any or all of which may appear in any given type. Since we mention a few of the specific eyepiece types later, we can talk about them more intelligently if we know what these defects are.

Spherical Aberration

This defect, which produces different appearances of the same star when the eyepiece is moved inside or outside the focal plane, appears not only in eyepieces but in the primary mirror as well. The effect of spherical aberration is the flaring of a star image when the eye is shifted off the axis of the eyepiece. It is less noticeable in dimmer objects but is always objectionable. It is an unfortunate characteristic of certain types of eyepieces and is especially bad at low powers when used with mirrors of moderate focal ratios—f/7 and below. It is caused by unequal distribution of light rays at the focal plane and produces difficulty in focusing. The eyepiece may be moved in or out of focus without changing the quality—or lack of quality—of the image.

Chromatic Aberration

The presence of color in the image indicates this particular defect in the eyepiece. The color appears throughout the field of view, but changes with the position of the eyepiece. If you observe an object of thin dimensions, such as a twig or a radio antenna, against a bright sky, it will appear

[1] For an excellent listing of various types see "Telescope Eyepieces" by Horace Selby in Albert G. Ingalls (editor), *Amateur Telescope Making,* Book III, Scientific American Publishing Company, New York.

to be outlined in red-orange when the eyepiece is inside focus but changes to shades of blue as the eyepiece is moved back. Strangely enough, if the same object is viewed against a dark background, the colors are reversed. Chromatic aberration is produced when the colored rays which together make up ordinary sunlight are bent unevenly as they pass through the various elements of the lens system. The most striking example is the separation of sunlight into its various colors as it passes through a prism.

Distortion

This is the defect which causes a straight line passing through the center of the field to become curved as it approaches the edges. The curvature is always convex toward the center of the field and becomes more pronounced as the distance from the center increases. Distortion is caused by unequal magnification of different parts of the image and is very similar to the wavy shapes seen at the corners of "wrap-around" windshields in automobiles. Fortunately this is not a very serious defect when looking at stars, but it would of course be intolerable in a telescope used for terrestrial observation.

Curvature of Field

This must not be confused with distortion. Its characteristic is lack of uniform sharpness throughout the field of view. If objects at the center of the field are in sharp focus, those at the edge will be blurred, and vice versa. The defect increases with the square of the distance from the center of the field. Users of microscopes are familiar with a difficulty similar to this one—focusing must be changed constantly to bring all objects in the field into sharp focus. The reason is the same in each case; the images are formed at unequal distances from the eye lens.

Oblique Astigmatism

Somewhat similar to curvature of field, oblique astigmatism occurs at an angle to the part of the image that is in sharp focus. If, for example, a particular pattern of leaves on a tree is seen clearly, another pattern at an angle to the first will be indistinct.

Chromatic Differences in Magnification

This is perhaps the most objectionable defect that can be found in an eyepiece. The center of the field is sharp and color-free, but as the eye progresses toward the side of the field, more and more color becomes apparent. These colors are of all hues and the general effect is that of looking through a prism. The difference arises from a difference in magnification of the elements of the spectrum. Each color is magnified, but some more than others. As in chromatic aberration, the color pattern is reversed depending upon whether one is looking at a bright object against a dark background, or vice versa.

TYPES OF EYEPIECES

Huygenian

Named for its inventor, the great astronomer whom Newton called "Summa Huygens," this eyepiece has some merits but, unfortunately, a greater number of deficiencies. It has an adequate field of view, fair eye relief, and there are no objectionable reflections (ghost images) within the eyepiece itself. It has many forms, but is usually made up of two plano-convex lenses of unequal focal length. The convex surfaces are turned away from the eye.

A feature of the Huygenian is that the focal plane lies between the two lenses, which precludes its use as a magnifying glass. It is called a "negative" lens for this reason.

Huygenians with the fewest errors have a field lens whose focal length is three times that of the eye lens, separated by a space one-half the sum of the two. Such an arrangement conforms to the formula given on page 148, and can be used to find the focal length of the arrangement.

The most serious defect of this eyepiece is a severe spherical aberration which is highly objectionable at moderate and low focal ratios, f/9 and below. It is also subject to distortion, curvature of field, and chromatic aberration at these focal ratios. But these defects become very much less noticeable at ratios of f/15 and above. For this reason it is widely used for long-focus refractors and reflectors.

On the whole, Huygenian eyepieces have so many more defects than virtues that they cannot be recommended for the kind of telescope we consider in this book.

Ramsden

Like the Huygenian, the Ramsden is made up of two plano-convex lenses, but in this case the curved surfaces face one another and are of equal focal length.

The Ramsden is a better eyepiece than the Huygenian but it suffers from many of the same defects, although in a lesser degree. Eye relief is short; it has a narrow field of view and considerable chromatic aberration. Ghost images are noticeable and so are distortion, curvature of field, and spherical aberration. These defects lessen as the distance between lenses approaches their common focal length, but in this case the eye relief is almost zero. Consequently the spacing is ordinarily reduced to two-thirds or three-fourths of the common focal length.

Errors in the Ramsden increase as the magnification decreases; hence it should not be used in short-focus telescopes.

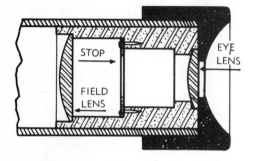

HUYGENIAN

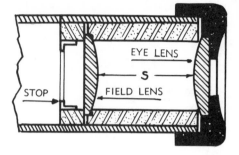

RAMSDEN

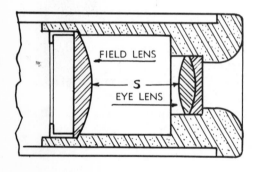

KELLNER

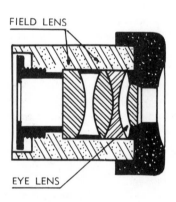

WIDE FIELD

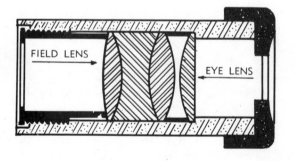

ORTHOSCOPIC

Types of eyepieces.

Shadows on the moon: Coper-
nicus and Mare Ibrium. Two-
second exposure on 2415 film.
(Photo by John Sanford)

Kellner (Achromatic Ramsden)

The Kellner is a modification of the Ramsden, the difference being that the plano-convex eye lens is replaced by a cemented achromat. This type of lens is made of two elements, one of crown glass, the other of flint glass, cemented together by some transparent substance such as Canada balsam. Each kind of glass has chromatic aberration, but when they are cemented together, one corrects the aberration of the other.

Kellners are excellent for low powers and are usually found in binoculars, field glasses, and spotting scopes. But they are also useful in medium-power reflectors where a moderately priced eyepiece is desirable.

Most of the defects we have seen in Huygenians and Ramsdens are either absent or greatly reduced in the Kellner. They have good eye relief, fields of view up to 50°, and very little distortion, curvature of field, or chromatic or spherical aberrations. An annoying defect, however, is the presence of ghost images which are especially noticeable when viewing bright objects. But if the lenses are coated with some nonreflecting material even this problem can be reduced to tolerable limits.

Solid (Monocentric)

These are eyepieces made from a single thick lens. There are two main types: the Tolles, which consists of the lens alone, and the Hastings, which makes use of another element cemented to the thick one. Because there are only two surfaces from which light can be reflected, these eyepieces give very bright images and are free of ghost images. They have good chromatic correction, sharpness of definition (especially at the center of the field), and very little spherical aberration. Some slight distortion is present, but the most important defect is a very small apparent field, only about 16°. One observer reports they give the impression of looking through a tunnel, especially after using a wide-field eyepiece. Although these eyepieces are of generally fine quality, they have been largely displaced by those listed below.

Orthoscopic, Plössl, König

These three eyepieces are superior examples of the eyepiece art. They are expensive but well worth their cost. They are characterized by flat fields and freedom from internal reflections, astigmatism, and distortion. With fields up to 50°, they have good eye relief and excellent contrast and color correction.

The orthoscopic design uses a cemented three-element field lens and a plano-convex eye lens. Depending upon focal length, they have apparent fields up to 60°, good eye relief, and a flat but contrasty field. They are excellent eyepieces for most telescopes, including the f/7 reflector.

Plössls use two identical double-element achromats which are set very close together. Eye relief is excellent, even better than the orthoscopic. Ghost images are negligible, the field of view is wide, and most of the other common eyepiece errors are almost nonexistent. They perform well in all telescopes and are notable for their pinpoint definition and image contrast.

The Königs are four-element eyepieces made up of a cemented achromat and two single-element lenses, although the arrangement may vary for different types. Not as well known as some of the others, still they are fine performers and are free from most eyepiece defects.

Edmund RKE

A patented eyepiece resulting from a computerized study, the RKE has a two-element achromatic field lens and a single double-convex eye lens. In effect this is a reversal of the Kellner, but is much superior to it. The RKE produces sharply defined images over the entire field. It has a stop placed at the focal plane which helps to create a sharp border to the image. Although it is much less expensive than the eyepieces mentioned above, its performance is on a par with any of them.

Wide Field

There are several types of wide-field eyepieces, most of which include a third element—usually an extra field lens—to create the wide field. One of them, the Erfle, uses an achromat for the field lens, then a double-convex intermediate single, and another achromat for the eye lens. One type uses three achromats. Although these are most useful in short-focus telescopes, they are often used as general-purpose eyepieces. Their eye relief is superior, which is a blessing for those who wear eyeglasses while observing, but the number of reflecting surfaces produces ghost images. The field of view can be as large as 80°, although 68° is the more common value. There is also some astigmatism and distortion at the edge of the field. The "giant" Erfle has a two-inch barrel and will not fit in the ordinary 1¼-inch eyepiece holder. This difficulty can be overcome by an adapter fitting, obtainable from almost any optical supply house.

The Erfle is a fine eyepiece for use with an RFT or Dobson telescope (see chapter 13) and yields breath-taking views of the Milky Way, expanded nebulae, comets, star clusters, and the like.

Zoom Eyepieces

Many users find this type a boon to general viewing. They have variable focal lengths—from 8 to 25 mm—accomplished by simply twisting the threaded barrels. Good ones remain in focus at any setting. In effect, they are the equivalent of several eyepieces in one barrel. They are expensive. An observer who can afford only one eyepiece will find this type invaluable.

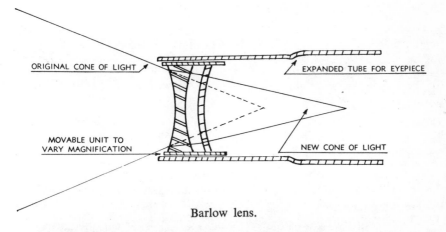

ORIGINAL CONE OF LIGHT

EXPANDED TUBE FOR EYEPIECE

MOVABLE UNIT TO VARY MAGNIFICATION

NEW CONE OF LIGHT

Barlow lens.

The Barlow Lens

Not, strictly speaking, an eyepiece at all, the Barlow is a lens combination placed between the diagonal and the eyepiece. It acts as a multiplier of the power of the eyepiece in use—up to 2½ times—depending on what Barlow is used. Most Barlows have fixed powers, but there is also a variable-power type in which the power can be changed simply by moving its lens system.

The Barlow lens is a negative lens which, when placed to intercept the original cone of light, narrows it down to a thinner and longer cone. The ratio of the two cone angles gives the magnification.

The Barlow is mounted in its own tube, which is expanded at one end to hold the eyepiece. The other end slips into the adapter tube of the telescope. A well-made Barlow, if used with an equally good eyepiece, presents several advantages. It increases not only the power but also the sharpness and brightness of the image. More important, it lengthens the eye relief. This is a great advantage when using short-focus eyepieces.

SOME SIMPLE LENS MATHEMATICS

As we have already seen, the magnifying power of a telescope may be determined in several ways. One of them is by dividing the focal length of the mirror by that of the eyepiece. Since eyepiece focal length is usually given in millimeters, it is best to use the same units for the mirror, as follows.

$$56 \text{ in.} \times 25.4 \text{ mm/inch} = 1422 \text{ mm}$$

Now suppose we have three eyepieces of 25 mm, 15 mm, and 8 mm. What magnification will each produce?

1	2	3
$1422/25 = 56.9$	$1422/15 = 94.8$	$1422/8 = 177.8$

What will be the size of the exit pupil for each? We use a simple formula:

$$\text{exit pupil} = \text{mirror diameter/magnification}$$

Since our 8-inch mirror equals 203 mm, we have

1	2	3
$203/56.9 = 3.57$ mm	$203/94.8 = 2.14$ mm	$203/177.8 = 1.14$ mm

Finally, what will be the eye relief? Assuming we know the back focal length (bfl) and using the information given on page 000, we can arrive at the following:

	1	2	3
bfl	24.06 mm	13.24 mm	7.8 mm
eye relief	24.5	13.4	7.85

Now suppose we use a 2X Barlow with each of the three eyepieces. Since the power is now doubled, the focal length must be halved, and we can derive the following table:

Focal length	12.5 mm	7.5 mm	4.0 mm
Magnification	113.8	189.5	355.6
Exit pupil	1.78	1.07	.57
Eye relief	11.0	7.4	3.7

The final two sets of figures represent the theoretical values for single eyepieces of the same focal length as the Barlow combination, but because the emergent cone is narrower and longer, eye relief will be larger than those given.

SOME PRACTICAL CONSIDERATIONS

The information given above tells us what we may expect from eyepieces of different magnifications, but what do we actually *see* when we use them? Let's

Jupiter. (Photo by
John Sanford)

consider some easily found objects and see what happens when we increase the magnification.

Let's assume that the planets Mars, Jupiter, and Saturn are in the sky at the same time, so we may shift from one to another quickly as we change eyepieces. We'll also assume that Mars is at opposition, when the planet is closest to us.

Using the 25 mm eyepiece we see a bright clear field. As we move from one planet to another, each is brilliant, small, and sharply defined, although little detail is apparent. Mars is just a glittering reddish disk. We see no markings on its surface, no polar caps. Jupiter shows us a little more. There seems to be a darkish line across the surface, and four bright specks of light are spread out in a line with the planet. We recognize the Galilean moons Io, Europa, Ganymede, and Callisto, although not necessarily in that order. This is a disappointment; Jupiter doesn't look anything like its NASA photographs. Saturn is even worse in this respect. All we can make out is a planet with "ears," much as Galileo himself must have seen it with his imperfect, low-powered telescope. But, after all, we are using only 56 power or 7 per inch of aperture.

Now we bring out the 15 mm eyepiece and at once things begin to improve. All the planets are a little less bright, but each occupies more of the field of view. Mars is still disappointing; there is no evidence of any detail on the surface, nor can we see a polar cap. But Jupiter looks much better. The single line across its face has turned into several ill-defined streaks, and the moons now appear farther away from the planetary disk than they were before. The planet is no longer round; it looks slightly ovoid. Saturn now shows dark gaps on either side, and the "ears" begin to look more like a ring.

The 8 mm eyepiece boosts the power to 175, or 22 per inch of aperture. The planets are larger, again a little less bright, and their edges are no longer sharp. The polar cap on Mars appears as a whitish indistinct little bump on the top of the planet. The surface looks mottled, although no distinct outlines are apparent. Jupiter's belts, however, are more distinct. We can see some bumps and wiggles in what had previously been straight lines. And we think we can see the famous Red Spot, not red at all but sort of an orange blob. The moons are still more distant from the planet. Their intensity varies and they seem to move a little, tiny variations in their positions. This is because we are magnifying not only the image of the planets, but also variations in the atmosphere through which we are looking. The ovoid shape of the disk is also more apparent. Saturn now looks the way we think it should. The rings are clearly apparent and we can even see indications of the Cassini division, even though it comes and goes as we watch. The surface of the planet is varied in color, but no detailed features are visible.

The 4 mm eyepiece jumps the magnification to 356, or 44 per inch of aperture. We are now approaching the limit of our telescope, and unless the atmosphere is free of turbulence and very transparent we are likely to be disappointed with what we can see through the eyepiece. True, the planets look larger, but the outlines of features seem to blend into the background. Focusing is difficult. Once in a while, though, if we are patient and persistent, we get a fleeting glimpse of detail. On Mars such moments show the south polar cap at the top of the planet, and we can see dark features such as Syrtis Major or Panderae Fretum. Jupiter reveals white splotches near the equator, the rings and zones are very apparent, and the orange oval of the Red Spot has more shape but less color than it did before. We begin to have a feeling for the immensity of Saturn's rings even though we see nothing of the detail of the NASA photographs. Although these glimpses are highly rewarding, we are almost forced to the conclusion that, on the whole, medium powers are more satisfying. When turbulence is bad, the high-powered eyepiece should remain in its container.

Does all the above apply to observing stars as well as planets? Decidedly not. The planets shine only by reflected light, which has to be spread over larger and larger areas as magnification increases. The stars produce their own light, which always appears as a point source. Nobody has ever seen the disk of a star, only the point of light it emits. Stars very close together—the binaries and those near the center of globular clusters—appear as single units to the naked eye or in low powers. But as magnification increases, they are resolved into their components, appearing ever more widely separated as magnification increases further. So this is where high power comes into its own. For example, the star Albireo at the foot of the Northern Cross looks like any other star with the naked eye. But when it is seen through a telescope it resolves itself into two individual stars, one a rich orange color and the other sparkling blue-green. Similarly, the globular cluster in Hercules is just visible to the bare eye as a small blob of light. Through high power it becomes a ball-shaped group of literally thousands of points of light, each an individual star.

WHAT EYEPIECES TO HAVE

We finally arrive at the most important consideration of all, a selection of eyepieces. We must base our choice on several factors: quality, cost, and ease of use. Quality is most important. Bargain eyepieces are no bargain at all; they waste the potential performance of our carefully made mirrors. It is best to buy from the selections of the established optical supply houses. Cost is a consideration important to everyone, and unfortunately prices rise constantly. The days when one could afford a battery of a dozen eyepieces are long since past for most of us. As a matter of fact, that many were never needed, and most of them sat unused in the eyepiece rack. For most purposes, three is an optimum number, either separately or two with a Barlow lens (which gives us the equivalent of four). Another possibility is a single zoom eyepiece. If possible, and within a reasonable price range, buy eyepieces that are threaded to take filters and are also parfocal, which means that they are designed so they can replace one another without changing focus.

The table below is included to give some estimate of projected costs, based on 1981–82 prices:

Ramsden	$ 7–9	Kellner	$16–20
RKE	30	König	30–45
Orthoscopic	25–50	Erfle	25–50
Plössl	30–45	Zoom	25–70

Barlow lens $25–45

Finally, we need an eyepiece selection which gives us a range of powers. Most amateurs find great enjoyment in expanded views of the heavens, and for this a low-power eyepiece is essential. For example, the Beehive open cluster in Cancer (Praesepe) is a magnificent sight when viewed in its entirety. It may be a disappointment when only a few of these jewel-like stars occupy the field, which is what happens when we use high power. The same is true for wide nebulae, comets, and other expanded objects. On the other hand, medium and higher powers are needed for viewing the moon and planets, where we want to see detail. We need the highest powers for separating the components of double stars, picking out individual stars in tightly packed globular clusters, and observing detail in faint objects such as the Crab Nebula.

It becomes obvious, then, that to exhaust the possibilities of the telescope we must have a range of power varying from low (25 mm) to high (8 mm). This should be the minimal equipment, but as time goes on most amateurs feel an itch to supplement it at each end of the scale.

MAKING THE EYEPIECE

Eyepiece making is a fascinating hobby. Those who become bitten by this particular bug find the hobby engrossing because it offers so many variations on the same theme and because the use of the end product is as exciting as its manufacture. There are even those who make telescopes to accommodate their eyepieces instead of the reverse process. To cover the whole field of eyepiece making would require another book the size of this one, so we shall confine ourselves to a practical discussion of making a simple eyepiece of low power.

You can, if you wish, do the whole job of making the eyepiece, from grinding and polishing the various lenses to mounting them in their cells for use in the telescope. Unfortunately, the actual manufacture of lenses requires a power-driven spindle and other special equipment. But you can have just as much fun if you skip this part and confine yourself to figuring out lens combinations from available sources and mounting them in cells of your own manufacture. You can find plano-convex lenses of the type you need from a multitude of sources. Old cameras, inexpensive pocket telescopes of the dime-store variety, view finders, field glasses, and other optical devices which have seen their best days are fruitful sources, provided the lenses are not chipped or scratched. Failing this, you can find several optical supply houses (see appendix) which have large stocks of war surplus lenses that are adequate if they meet your requirements.

For equipment, a lathe is very helpful but not essential. You will need some files, a hacksaw, a sharp chisel or two, and some airplane cement. You can buy 1¼-inch O.D. (outside diameter) thin-walled brass tubing, nickeled on the outside, from almost any plumbing supply house or hardware store. Plastic tubing—the Bakelite type—of varying wall thicknesses to fit inside the brass tubing can usually be obtained from the same sources. If not, plastic tubing of almost any dimensions can be bought direct from the manufacturer. Finally, you need a few small pieces of flat plastic stock —Bakelite or any other kind—of different thicknesses.

How to Find the Focal Length of a Lens

Because focal lengths are rarely marked on lenses, you must check each one to be sure it is the focal length you desire. This is not a difficult task. Mount a piece of screening over a ½-inch hole cut in a piece of white cardboard the size of a playing card. Mount the cardboard in a vertical position near one end of a yardstick or meterstick. You can do this by using a small piece of modeling clay stuck to the upper edge of the yardstick. Now place a low-wattage electric light bulb, such as the small baseboard fixtures used for night lights, directly behind the screening, that is, toward the short end of the yardstick. Now hold another small piece of cardboard in one hand and the lens to be tested in the other (see diagram), and move both cardboard and lens to various positions on the yardstick until you get the image of the screening sharply focused on the cardboard. Note the approximate positions of each on the scale of the yardstick. Place a small piece of modeling clay on each of these positions. Stick the lens in one piece of the modeling clay, then place the cardboard in various positions in the other piece until you find the point at which the image of the screening has the sharpest possible focus. Be sure that all three elements, screening, lens, and cardboard, are each perpendicular to the yardstick.

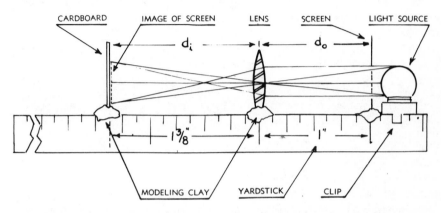

Measuring focal length by means of a simple optical bench.

The distance from the screening to the lens is called the object distance, or d_o, and that from the lens to the cardboard is known as image distance, or d_i. Measure these distances carefully, using the scale on the yardstick. Now you can calculate the focal length of the lens, using the formula

$$f = \frac{d_i \times d_o}{d_i + d_o}$$

Using an example from the diagram, if $d_i = 1\frac{3}{8}$ inches and $d_o = 1$ inch, then

$$f = \frac{1\frac{3}{8} \times 1}{1\frac{3}{8} + 1} = .58 \text{ inch} = 14.7 \text{ mm}$$

Two lenses of the focal length just found will produce a short focal length Ramsden which, when used with your 56-inch focal length mirror, will yield a magnification of 129 and an exit pupil of 1.57 mm.

Now let us follow the steps by which we arrive at these figures.

1. Finding the *efl*, using two lenses of 14.7-mm focal length. From the formula on page 148, and remembering that the spacing must be $\frac{2}{3}$ of the focal length of either lens, as mentioned on page 161, we get:

$$efl = \frac{14.7 \times 14.7}{14.7 + 14.7 - \frac{2}{3} \times 14.7} = 11.0 \text{ mm}$$

2. But since we have the *efl* in millimeters, we must change all other dimensions to millimeters also.

$$\text{Magnification} = \frac{\text{focal length of mirror}}{\text{focal length of eyepiece}}$$

$$= \frac{56 \text{ inch} \times 25.4 \text{ mm/inch}}{11 \text{ mm}} = 129$$

3. Finally, since exit pupil equals diameter of mirror divided by magnification (page 151), we have

$$\text{exit pupil} = \frac{8 \text{ inch} \times 25.4 \text{ mm/inch}}{129} = 1.57 \text{ mm}$$

Now let's assume that you have two lenses of the above focal length and wish to make an eyepiece from them. This is a special example, of course, but you can figure out the dimensions for *any* pair of lenses using the above formula.

Making the Body of the Eyepiece

Start by cutting a piece of $1\frac{1}{4}$-inch O.D. brass tubing $1\frac{1}{4}$ inches long. Wrap a piece of paper around it close to an end of the tubing so that one edge of the paper lines up straight with the wrapped edges of the paper underneath. Using this straight line as a guide, file the end of the tubing until it is square with the length. Repeat at the other end. The inside of this tubing must be black so that it will not reflect light. Wrap the outside of the tube with paper, then spray the inside with flat black paint from a spray can that you buy in a hardware store. Buy a large can since you will neeed it later to spray the inside of your telescope tube.

While you are waiting for the paint to dry, mount the lenses with convex faces toward each other on a yardstick, using modeling clay again for support. Place them at a distance equal to their common focal length, or 14.7 mm in the example. Call the lens nearest the end of the yardstick the eye lens, and look through it toward the field lens. If the lenses are exactly their common focal length apart, you will see dust marks, fingerprints, and other marks sharply outlined on the field lens. This would, of course, be intolerable in an eyepiece since every little dust speck on the field lens would interfere with your vision. But if you move the two lenses closer together so that the distance between them is about $\frac{2}{3}$ their common focal length, you find that the dust specks and fingerprints disappear. This is because they are now out of focus as far as the eye lens is concerned and therefore invisible. Be sure that the axes of both lenses coincide with each other—that is, that the lenses are parallel to each other and that their centers are on the same line. Measure the distance between the convex surfaces carefully since this is the distance at which you will place them in the mounting. Call this distance *s;* it should be close to 10 mm.

Place the wire screening in front of the field lens and move it back and forth until it is sharply focused as you look at it through the combination of the two lenses. This is the distance from the field lens at which you will place the field stop or diaphragm in the eyepiece tube. Call this distance d.

Cut a piece of plastic tubing of the length s found above. (The $\frac{2}{3}$ focal length is generous enough to allow for convexity and thickness of lenses.) Square the ends of the tube carefully, blacken the insides, and cement the lenses to each end, convex surfaces facing each other. A drop or two of airplane cement on the *rims* of the lenses will do this nicely. Be careful that you don't smear any of the glue on the lens surfaces; if you do, remove it with acetone or amyl acetate. Also be very sure that the inner surfaces of the lenses are perfectly clean before attaching them to the tube. Check again to see that the lenses are lined up with each other.

After the cement is thoroughly dry, slide the plastic tubing containing the lenses into the brass tubing. The eye lens should be flush with one end of the tubing. You can cement the tubes together, in which case you won't be able to get them apart again for cleaning, or you can secure them by using countersunk spectacle screws or $\frac{1}{16}$-inch Allen set screws. If you adopt the latter course you will have to drill and tap the brass tubing for the screws.

Cut off a thin slice of the plastic tubing exactly the length of dimension d measured above. This slice will serve a double purpose. It holds the field lens securely in place and is also a spacer for the placing of the field lens stop or diaphragm. Square the faces of the spacer with a file and cement it into place against the field lens, after blackening its inner surface.

Now cut a disk from $\frac{1}{16}$-inch Bakelite sheet exactly the inside diameter of the brass tubing, drill a $\frac{7}{16}$-inch hole in its center, and clean up the inside edges with a round file. Cutting this is a difficult operation unless you have a lathe, but it can be done without one. Cutting this is a simple operation with a lathe, but it can be done in two other ways. Cut out the disk with a coping or spiral-blade saw to a trifle more than the inside diameter of the brass tubing. Drill a $\frac{3}{16}$- or $\frac{1}{4}$-inch hole in the center of the disk, insert a bolt to fit snugly, screw a nut down tight against the disk, and put the bolt in the chuck of your drill mounted on a bench grinder. Then, with the drill turning the disk, smooth down the edge of the disk with a piece of Carbo stone until it just fits the tubing. After this, enlarge the hole to $\frac{7}{16}$ inch.

Or you can use the biscuit-cutter arrangement described on page 144. Make one the size of your eyepiece tubing and cut out the circle. If the finished product does not fit the inside of the tube perfectly, no harm is done if you are sure to line up the hole in its center with the axis of your lenses. Cement it into place against the spacer ring you have just inserted.

A lathe is also useful in the final operation, making the eye cap. Obtain a piece of Bakelite or other plastic ½ inch thick and cut a circle 1½ inches wide. If a lathe isn't available, use a biscuit cutter for this operation. Drill a ⅜-inch hole in the center. Now, using another biscuit cutter, cut a circular ring in this disk ¼ inch deep and the size of the outside diameter of your eyepiece tube, or 1¼ inches. Clamp the disk in a vise and hollow out the center. A small sharp woodworker's chisel will do this job. Sandpaper the depression until it is smooth and clean. The hollow you have made should fit tightly over the outside of the eyepiece tube in a "push" fit. If it does, make a tapered hollow on the reverse or eye side of the disk about ⅛ inch deep for eye relief. Round off the edges, push the cap on the eye lens end of the brass tubing, and you have completed your eyepiece. Of course this whole operation is much simpler if you are fortunate enough to possess a lathe.

The eyepiece you have just made will have all the faults of any Ramsden eyepiece. Compared to an orthoscopic of the same power, its performance, you may find, leaves something to be desired. On the other hand, you may be so delighted with your handiwork that you will want to make other and more complex eyepieces. The manufacture of these types is beyond the scope of this book, but there is voluminous literature on the subject.[3]

With the completion of mirror, diagonal, and eyepiece, you have completed the optical parts of your telescope. All that remains to be done now is to construct the fittings which will hold these optics in their proper position in the telescope tube. This is an operation of fundamental importance because good optics require the best and sturdiest support. You should keep this in mind as you read the next chapter.

---

[3] See Louis Bell, *The Telescope,* McGraw-Hill Book Company, New York (out of print but available at libraries); A. E. Conrady, *Applied Optics and Optical Design,* Oxford University Press, New York (out of print but available at libraries); "Simple Oculars" by Russell W. Porter, R. E. Clark, and W. T. Patterson in Albert G. Ingalls (editor), *Amateur Telescope Making,* Book II, Scientific American Publishing Company, New York; "Telescope Eyepieces" by Horace Selby in *Amateur Telescope Making,* Book III.

CHAPTER 12

The Telescope Tube and Its Fittings

THE TELESCOPE TUBE

THERE ARE MANY variations in the kind of tube which can be used for an 8-inch telescope. The tube may be round, square, many-sided, or open (that is, simply a framework for supporting the optical elements). It may be constructed of steel, aluminum, wood, fiberglass, or even heavy cardboard. Choice of shape or material is a matter of individual preference as long as the maker incorporates one principal characteristic in his tube—rigidity. Telescope mirrors and the accessories which support them are fairly heavy and must be supported in a structure that permits no bending or flexure of any kind. The ideal tube is one which incorporates strength and lightness, especially if the telescope maker hopes to produce a portable instrument.

Fabricating steel or aluminum into tube form is a difficult task; hence, if you choose either of these materials for your tube, it is better to have this done at a machine shop where the proper equipment is available. You can buy your tube ready-made, if you wish, for there are many sources of supply for this kind of equipment. A good aluminium or steel tube for an 8-inch mirror costs about $35. But metal tubes are affected by changes in temperature; they expand and contract with the weather and can seriously affect the performance of the telescope. Much better and cheaper are the paper construction tubes used for concrete forms in the building industry. They are strong and rigid, have good insulating characteristics, and when painted look neat and clean. They can be bought by the foot and come in sizes of 2-inch variation in diameter. Most have walls ⅜ inch thick or more.

Wood tubes are strong and light. Many amateurs make them with square cross sections, but in this form they are awkward to use because the eyepiece is sometimes difficult to reach unless the owner is a gymnast. More difficult to make but easier to use because they can be rotated in the mounting saddle are multisided wooden tubes, octagons, or dedecahedrons (twelve-sided). These are assembled from clear pine strips glued together. The edges of each strip must be cut at an angle so they fit tightly together. For an octagonal tube the angle is 22.5°, for the twelve-sided it is 15°. If the wood strips are thick enough, the edges may be planed or sanded to make a nearly round tube. Other materials may also be used in this way. One amateur built his of Gatorfoam and says it is light, rigid, and solid.

The material which best combines lightness with rigidity is fiberglass. It has a specific gravity of 1.8, compared with 2.7 for aluminum, which means it is only $\frac{2}{3}$ as heavy, volume for volume, as aluminum. It will not lose its shape, dent, or bend, needs no painting or upkeep, and is absolutely corrosion-proof. It may be sawed, riveted, machined, drilled, and tapped. But perhaps its outstanding characteristic as a material for telescope tubes is its low thermal conductivity. Like any glass product, it expands and contracts very little with change of temperature, a failing of all metal tubes. This is important in keeping the optical elements aligned, and it also cuts down on convection currents within the tube itself. Finally, it is unaffected by change in humidity.

There are several suppliers of fiberglass tubes for telescopes. Although the cost is a little more for fiberglass than for other materials ($50 for a tube for an 8-inch mirror as against $30–$35 for metal tubes of the same dimensions), they are well worth the extra cost. If carefully built up of reinforced plastic, they are stronger, pound for pound, than any other tube type. If you decide to order fiberglass, be sure that the manufacturer specifies that his tube is reinforced. Some tube makers simply cover paper tubes with a thin fiberglass coating. This is satisfactory for a small tube, but a large one needs more "bones" for real support. .

Although the task of constructing tubes of this type might seem beyond the abilities of the amateur, actually it is much easier than it appears at first glance. And it is the most inexpensive tube, with the exception of wood, that it is possible to make. The one shown in the illustration costs $18, including the materials for making the mold upon which it was built. The job is a messy one and the odors from the resin are somewhat overpowering, so it must be done in a well-ventilated room. But aside from these minor difficulties there are no others that the determined tube maker can't overcome. You must make a cylindrical mold whose outside diameter is the same as the inside diameter of the proposed tube. You build up a laminate of polyester resin as the bonding agent. Made this way, your finished tube will have a wall thickness close to $\frac{1}{4}$ inch and will be light, strong, and tough.

You need the following supplies:

A collapsible mold made in two parts so that it can be easily removed after the tube has hardened (Make this yourself [see below].)

A supply of newspapers or building paper

A piece of heavy polyvinyl plastic sheet (at least 10 mil size), 30×60 inches (It is a good idea to buy an extra yard or two in case of difficulty.)

A roll of 1-inch masking tape

2 quarts of polyester resin (This comes under various trade names, but you should get the type which is made for vertical surfaces.)

4 oz. resin thickener

¼ oz. accelerator (cobalt naphthenate)

2 oz. catalyst (methyl ethyl ketone peroxide)

A piece of fiberglass mat, ⅛×30×60 inches

3 yards of fiberglass tape, 3 inches wide

2 yards of fiberglass cloth, heavy, 36 inches wide

Coloring matter for the resin

A 3-inch paint brush or roller

A window-cleaning squeegee (from the dime store)

Rubber gloves

1 quart of polyester resin solvent, or some commercial acetone, for cleaning hands, paint brushes, etc.

You can buy all these from boat yards, mail-order houses, or several suppliers which make a specialty of fiberglass products. There is a list of supply houses in the appendix.

Making the Mold

There are two ways to do this. The simplest is to find a heavy cardboard tube, the kind upon which carpets are rolled or the heavy tubes

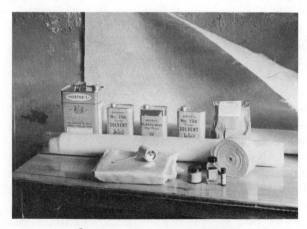

Raw materials for making a fiberglass tube. The transparent material at left is the vinyl plastic used to line the tube.

used as forms for the construction of concrete posts. As noted above, the outside diameter of the tube should be equal to the inside diameter of the telescope tube you wish to make. For your 8-inch mirror, you will want at least a 9-inch I.D. telescope tube, and an extra half inch added to this will do no harm. Make your tube an inch or two longer than the focal length of your mirror—you will not need this extra length for construction purposes, but it helps keep extraneous light from your diagonal mirror.

Slice the tube down the center by cutting out a strip ½ inch wide on opposite sides. From a piece of 1×8 or 2×8, cut out eight half-moons of a radius to fit the inside of the cardboard tube, which you will tack at equal intervals down the inside of each of your half cylinders. Cut a "V" ½ inch deep in the center of each half-moon. Place the two halves of the mold together, separating them ½ inch by means of a pair of wooden strips of this size placed lengthwise and secured in place with strips of masking tape. Tack two cleats across each end of the mold, leaving space between them so you can insert a piece of half-inch water pipe down the center of the mold. Remove the temporary spreaders, slide the length of water pipe down the center (this is the reason for the "V" cuts in the half-moons), and you are ready to use the mold. Support the ends of the water pipe in a couple of pedestal chocks mounted on your workbench. The mold will now revolve on the water pipe and you can turn it easily as you add the fiberglass components to the outside.

If you can't find a tube the right size for a mold, you can make your own out of wood. Cut out four 7½-inch circles from a piece of 1×8 and convert them into the half-moons mentioned above by sawing a half-inch strip out of the center of each. Place four of them at regular intervals on your workbench so the total length is 60 inches. Nail beveled 1×2-inch strips along the half-moons as shown in the photograph, placing the strips close together. If you don't possess a table saw with which you can rip out the bevels, there will be gaps between each strip, but this is not of vital importance since you will cover the mold with heavy paper anyway. However, beveling each strip 10° makes a much neater job. Sand the projecting edges of the strips down until each of the half polyhedrons you have made becomes a half-cylinder. You will find that you can sand the crude apparatus into a very close approximation of a smooth-curved cylinder. Check the curve with a template sawed out of a piece of ¼-inch plywood. If, as a result of your sanding, you end up with a cylinder whose diameter is lightly less than you had hoped for, don't worry since you can build it up by adding a sufficient amount of newspaper or building paper to its surface. Put the two halves together according to the directions for the paper tube above.

A wooden mold made in this way is much more satisfactory than the one made from cardboard since it is more rigid and will stand more abuse. To prevent warping, shellac it inside and out with two coats of orange shellac. Once you possess a good mold of this type you can manufacture tubes for your telescope-making friends, which will spread out the original cost ($5 to $10). The mold shown in the illustration was completed in three hours, with access to a power saw.

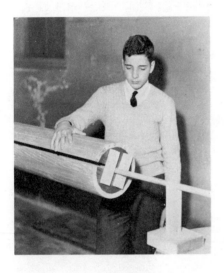

How the mold for the tube is set up. The longitudinal spreaders have not been inserted.

Building Up the Tube

Wrap the mold in several layers of newspaper or building paper. This is to protect the inside of the tube from becoming scarred when you remove the mold after the tube has been built up.

Polyvinyl plastic sheet can usually be obtained at hardware or department stores. If you can obtain plastic which has a flat black finish, so much the better. If not, buy any color you like, tack the plastic to a flat surface, and spray one side with flat black from a spray can. Do this carefully and evenly, for this will be the inside of your telescope tube. It is much simpler to paint a flat surface than to have to paint the inside of a tube.

Wrap the plastic around the mold with the black side in. Cut it at a point which will produce a half-inch overlap along the length of the tube and trim the ends about four inches longer than the focal length of your mirror. Cut longitudinally through both layers of the overlap with a sharp blade. Butt the ends together and secure with a strip of Scotch Tape. This makes a seam which is almost invisible.

Now mix the coloring matter carefully with the polyester resin. Stir with a wooden paddle until the color in the whole batch is uniform. Polyester resin usually requires two additional compounds, one which causes the resin to harden and the other which speeds up the process. Different manufacturers treat the addition of these compounds in various ways, but all give explicit directions for their use. Follow these directions exactly. The addition of just a tiny amount too much of either hardening compound (accelerator) or setting compound (catalyst) speeds up the process far out of proportion to the amount added. You will find yourself facing a sloppy, sticky mass if you add too little, or a containerful of already hard plastic if you add too much. There is one characteristic of this type of plastic material which you should remember, however. Once you have added the setting compounds to the resin, the plastic will harden in a period determined by how much was added. There is no way of reversing the process. It is thus very important that you mix up only small batches at a time. If you haven't made enough for a particular operation, you can always stir up some more. This restriction applies only to the setting compounds; it has nothing to do with coloring matter or thickening material. These may be added at any time and to any amount of the plastic.

Having prepared the resin according to the manufacturer's directions, apply a thick coat to the polyvinyl plastic covering the mold. Allow it to dry until it becomes tacky. While you are waiting, cut the fiberglass mat into a sheet the same length as the plastic sheet, and about 29 inches wide (or whatever width will just equal the circumference of the tube with the plastic on it). Place this on the mold, stretch and smooth it until the meeting edges are flush with one another, and sew along the seam with an overlap stitch, using thin nylon thread. Some workers prefer to have the fiberglass mat overlap along the seam, but the resulting bulge will have to be sanded off later.

Now wrap the tube in a piece of fiberglass cloth cut to the same dimensions as the mat. Smooth this out very carefully, pulling and stretching until the cloth makes a tight fit over the mat. Sew the edges together as you did with the mat.

The whole tube must now be *saturated* with resin. In order to prevent the resin from running or dripping, it is a good idea to add a little thickening agent to it. Most manufacturers supply this along with the resin or sell it at small added cost. If used according to their directions, it makes the addition of the resin to the fiberglass cloth and mat much easier to accomplish. Apply liberally with a paint brush or roller. Work the resin into the surface with a squeegee or your fingers or both. The resin must saturate both the fiberglass cloth and the underlying layer of fiberglass mat. This is the key operation in the manufacture of the tube. Unless both fiberglass mat and cloth are *thoroughly* saturated with resin, large bubbles will form between them in spite of all your efforts, the mat will not bond to the polyvinyl plastic underneath, and you will end up with a lumpy tube which is anything but beautiful. Do only short sections of the tube at a time. If the resin starts to harden before you have applied all of it to the tube, dump the remainder and mix a new batch. You can tell that the plastic is about to harden by an increase in temperature. During this operation small bubbles will form in the resin. These must be worked out with the fingers or they will leave holes when the resin dries. Smooth the surface as evenly as possible.

Allow the tube to harden, or "cure," for at least 24 hours. Sand off any rough patches on the surface, then apply another coat of resin, again being very careful to get rid of any bubbles which form. If, in spite of all your precautions, there are still a few small bubbles present after the resin has set, sand these lightly, allowing the sanded resin to fill in the holes. The next coat of resin will seal the holes solidly.

Apply two more coats of resin, smoothing the surface carefully each time. Sand after each coat. With additional coats of resin and careful sanding, a surface may be built up which is as hard and as smooth as glass. You will decide for yourself when you have reached the point where further applications will not improve your tube.

To remove the mold, knock off the two cleats at each end and carefully loosen the two half cylinders. If you take your time, you can remove the mold without disturbing the paper between the mold and the inside of the telescope tube. Now remove the paper lining just as carefully as you did the mold. This should present no difficulty, although there is the possibility that some of the resin may have leaked through the seam in the polyvinyl plastic layer on the inside. If this has happened, tear the paper away from the spots where it is sticking, then swab gently with a little solvent. The paper will loosen and come away without marring the inside of the tube.

A 9½-inch fiberglass tube and the mold from which it was made. The sticks leaning against the mold are the longitudinal spreaders used so that the sections may easily be separated.

The tube should be completely tough and solid, but if there is any sign of flexure when it is subjected to pressure, add another layer of fiberglass cloth and finish off with resin. You need not replace the mold to do this, although you may decide to do so for convenience in handling. But it is more than likely that you will be completely satisfied with the tube and will find any further treatment unnecessary. There is one other addition you may wish to make, however. For appearance' sake, it is a good idea to build up the ends of the tube until they *appear* a little stronger. You can do this by adding a few layers of 3-inch tape to each end (or whatever width suits you) and saturating with resin. Some workers prefer to do this with resin of a contrasting color.

THE MIRROR CELL

The cell that supports the main mirror must be designed to hold it firmly in position but at the same time impose as little restraint upon it as possible. The simplest way of accomplishing this is to provide three points of support near the edge of the mirror, spaced 120° apart. These supports must be adjustable to make it possible to change the position of the mirror easily.

There is some variance in theory and practice as to where the mirror's supports should be placed to provide for even distribution of weight. It can be shown mathematically that 3 points spaced evenly at .7 of the radius from the mirror's center satisfy this condition. The great majority of mirror makers use this system. On the other hand, it is claimed that the mirror has a tendency to sag away from its supports, thus producing astigmatism. Because of such sagging, the argument goes, it is better to place the supports near the edge of the mirror. The droop in the glass merely deepens the curve a little, a condition which can be compensated for in focusing. While there is no doubt that a certain amount of astigmatism is introduced by the .7 radius system, it is questionable that it is serious enough to impair the definition of the mirror. When the supports are shifted to the edge, new unknowns—how much the mirror will sag, and whether it will sag evenly—are introduced. So we shall balance the mirror on three points spaced 120° apart, each 2.8 inches from the center, and take our chances with the astigmatism thus introduced.

It might be supposed from the above that the most efficient way to support a mirror would be to cement it to a very solid and unyielding base. This is without question the *worst* possible way of doing it, for two reasons. When any surface rests upon another, the two surfaces make contact at only three points, no matter how flat the surfaces may *seem* to be. This is not conjecture; it is a physical fact. But there is no way of being sure where these three points of contact will be when we have one surface simply resting upon another, and it is entirely possible that they may be in the most unfavorable positions from the viewpoint of providing adequate support. Cementing the surfaces together does not take care of the problem, it simply changes the position of the supporting points. Furthermore, cementing the surfaces produces another bad effect, which comes from the unequal coefficient of expansion of the glass and the material to which it is cemented. Changes in temperature thus produce a strain in the glass which in turn results in a deformation of the mirror surface.

Having decided upon the supporting system for the mirror, we must now find a way to hold it on these supports. A simple way is to place

Open-cell design for a 12-inch mirror. It is made from cast aluminum and is lighter than it looks in this picture.

a metal ring around the glass, then provide three pins lateral to the axis of the mirror, each faced with cork or rubber against which the sides of the mirror may rest. These pins are also placed 120° apart. The mirror will rest against only two of them at any one time. This is very important, for if all three press against the glass tightly enough to hold the mirror immovable, they also produce a strain in the glass. This strain, of course, varies with the temperature, for the supporting ring and the mirror will not expand and contract at the same rate. Strain in a mirror causes astigmatism and a consequent lowering in performance. It is very important, therefore, that the lateral supports press against the mirror only enough to hold it *firmly* but not with an unyielding grip. We once had a striking example of this at the Millbrook Observatory. Our 12½-inch mirror, which had always been reliable, suddenly began to misbehave and produce images of poor quality. We checked everything we could think of—collimation, diagonal, eyepieces—and finally discovered that one of the boys who took care of the observatory had tightened the lateral adjustments of the mirror as tightly as he could turn them for fear that the mirror might fall out of its cell. When the pins were loosened to their normal position, the mirror once again behaved well.

In order to prevent accidents when the telescope tube is tipped downward, clips are placed at the edge of the mirror. These clips touch the glass very lightly and, like the lateral ones, should be faced with a yielding material. Felt strips glued to the tabs are useful for this. They bear against the bevel on the *edge* of the mirror, not against the mirror's surface. They are placed in such a manner that if the mirror tips forward it cannot fall out of its cell. Usually there are three of them, located at 120° intervals around the rim.

Making a Simple Mirror Cell

Cut two 8½-inch circles of good quality plywood, ½ inch thick. Glue them together, using a casein glue or any other adhesive designed for wood. Sand the surfaces and the edges, then apply three coats of orange shellac and two coats of varnish to seal the surface.

Find the center of the disk and from it draw a circle of 2.8-inch radius. Keeping the compass set at this radius, strike off six successive arcs around the circle. If you do this carefully, the final arc should end at the point where the first arc was started. Each of these arcs is, of course, 60° apart; using alternate ones you will establish three points 120° apart.

Drill a ³⁄₁₆-inch hole at each of the three points. Countersink the nuts from three ¼×1½-inch galvanized bolts in each hole. Grease the bolts and screw them through the holes from the back of the cell until they pass through the countersunk nuts and project about ¼ inch above the surface. File the bolt ends to roundness.

Have a blacksmith make you an 8½-inch I.D. ring with a collar 1½ inches high out of ⅛-inch soft iron. Drill this at several places around the ring and screw it to the same side of the wood base as the rounded bolt ends. Tap and drill the collar at three points 120° apart for ⅛-inch Allen set screws faced with cork or rubber on the ends which bear against the mirror. After placing the mirror in the cell, turn the set screws in until the mirror is centered and is making light contact with all three of the set screws.

Measure the amount the mirror projects above the collar so that brass tabs may be placed around it as a safety device. Remove the mirror, drill and tap three more holes spaced between those for the lateral positioning Allen screws, and attach slotted tabs bent far enough inward

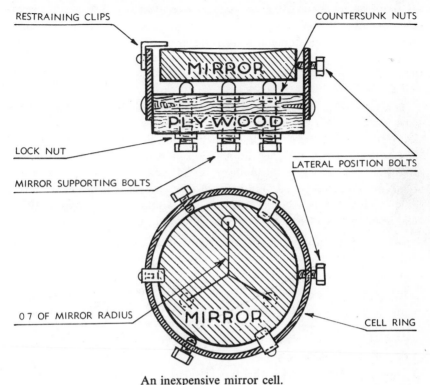

RESTRAINING CLIPS

COUNTERSUNK NUTS

MIRROR

PLYWOOD

LOCK NUT

MIRROR SUPPORTING BOLTS

LATERAL POSITION BOLTS

0 7 OF MIRROR RADIUS

MIRROR

CELL RING

An inexpensive mirror cell.

to bear lightly on the bevel at the mirror's edge. Line the inside of each with felt. Paint all inside surfaces of the cell with flat black.

Slide the cell (without mirror) into the end of the telescope tube. The outside diameter of the cell as described above is 8¾ inches, while the inside diameter of the tube is between 9 and 9½ inches, depending on how you made your mold. You can fill up the extra space by attaching three pieces of flat iron of the necessary thickness, 2 inches long and ½ inch wide, to the inside of the tube. The sides of the cell will rest against these supports. You could, of course, build up the inside of the tube with fiberglass mat and resin, thus producing an exact fit for the cell inside the tube. In any case, secure the cell in place by means of three set screws placed at the end of the tube. With the cell in place, drill through the tube and the rim of the cell. Since the hole in the rim will be inside the tube, you must locate its position for future use by placing some kind of mark on the base or side of the cell. This is to be sure that you can line up the holes easily whenever you wish to remove and replace the cell in the tube.

The above is a simple and inexpensive cell which will hold your mirror securely and which permits adjustment of the mirror position without removing it from the tube. You will find this a great advantage when the time comes for you to line up, or collimate, the optical elements of the telescope.

A Fiberglass Cell

You can make a more professional-looking cell by means of a fiberglass casting. Very few amateurs have attempted this, so if you wish to pioneer in one aspect of telescope making, here is your opportunity. The idea comes from the Long Island telescope maker, Allan Mackintosh. Like so many of the other contributions he has made to the field, this is a sound one. It is suggested here because it is an excellent means of mounting the mirror and also because it is an addition in keeping with the tube itself. You can make it from the following materials:

> Wooden patterns of the cell parts you wish to duplicate in fiberglass (It is better to make the parts individually than to attempt to make the whole cell in a single casting.)
> Five pounds of plaster of Paris or dental cement (The latter is more expensive but makes much better molds for castings.)
> A quart of polyester resin and a few ounces of thickener
> A pint of mold release compound
> A cake of paraffin
> A yard of heavy fiberglass cloth (10 oz.)
> A strip of polyvinyl sheet plastic, 3×30 inches

The Pattern for the Casting

You can make any type of cell you like as long as it is rugged enough to support the mirror adequately. The one suggested here is given only as an example; your own ideas may very well be superior to the ones presented.

Fiberglass $\frac{1}{4}$ inch thick has strength superior to an equal thickness of aluminum, so we shall consider this dimension as adequate for our purpose. Ventilated cells—that is, those which permit a flow of air around the mirror—are better than closed ones, especially if you have a fiberglass tube where convection currents are cut to a minimum. For a trial design, then, a three-pronged base surrounded by a circle is a safe way to start.

From a piece of $\frac{1}{4}$-inch plywood cut a circle whose diameter is the same as the inside diameter of your telescope tube. See illustration at bottom of page 179. Make each arm and curve 1 inch wide and the central boss twice this width. Cut off four lengths of broom handle or 1-inch dowel, each $\frac{1}{4}$ inch long. Glue these to the top of the central arms as shown. The centers of the pieces of dowel must be 2.8 inches from the center of the pattern. Glue the fourth disk to the center of the central boss. Now cut out three lengths of $\frac{1}{4} \times \frac{1}{4}$ -inch pine and glue them into place between the dowel pieces as shown.

Using wood rasp and sandpaper, round off all surfaces except the outside edge of the disk, which should be sanded with square edges. Sand the whole pattern smooth and cover with two coats of orange shellac and one coat of varnish. Fill in all surfaces which meet at right angles with paraffin or fillet cord and smooth them until the right angle has been changed to a rounded curve. A $\frac{1}{4}$-inch ball bearing, warmed until hot to the touch, rolled along each of the joints, will do this job admirably. Rounding these surfaces serves two purposes; it makes the finished product look better and also prevents the pattern and casting from sticking to the mold.

A flat cake pan large enough to accommodate your pattern serves as a good container for the plaster of Paris. Grease this container and the pattern. Make up a batch of plaster of Paris or dental cement in the pan and press the pattern into it, reinforced side down, until the surface of the plaster is flush with that of the pattern. Hold the pattern steady until the plaster sets (about 10 minutes). Allow it to dry for 24 hours before removing the pattern. Use a little patience in getting the pattern out of the plaster; rough treatment can be disastrous. Patch up any chips, cracks, or holes in the mold with plaster of Paris paste. Allow the mold to set for at least 48 hours or, if you are in a hurry, bake the mold in the kitchen oven for several hours at a temperature of about 150° F. After allowing it to cool, paint the inside of the mold with five or six coats of mold release, allowing each to dry before applying the next.

Now cut out two patterns of fiberglass cloth exactly the dimensions of the wooden pattern. Cut some more fiberglass cloth into strips, circles, or other shapes which will fit various parts of the mold. This is easier than attempting to cut out a series of whole patterns, for they are very hard to handle in the casting process, especially ones which would have to be placed near the bottom of the mold.

Paint the inside of the mold with a thick layer of resin to which you have added the recommended amount of filler (see manufacturer's directions for this), add a layer of pieces of fiberglass cloth, then another thick resin coat, and so on. When the mold is filled to the level of the reinforcing bars, add a complete pattern of cloth, and add another one sometime later on before you finish the job. All of this sounds difficult, but the job goes surprisingly fast and you should complete it within 45 minutes. It is better to make two batches of resin for this job, stopping in the middle to make the second one. In this way you can add a little more of the setting compounds than you ordinarily would, which makes the casting harden more quickly and also makes the finished casting easier to get out of the mold. But in your eagerness to finish the job, don't add too much catalyst. The plastic, if it sets too fast, gives off considerable heat, which will cause your casting to crack. Your casting, in places, will be $\frac{1}{2}$ inch thick and takes a correspondingly longer time to set than the thin layers of resin you used on the telescope tube.

When you have finished, put the casting and mold away in a warm, dry place and allow it 48 hours to harden. You may find it impossible to remove the casting from the mold at the end of this time in spite of all your care. If this happens, break the mold carefully; you can always make a new one if you decide to make another cell later on. Your casting should be smooth and in good condition when removed from the mold, but if there are rough spots you can clean them up with file and sandpaper.

You now wish to make a rim for the base you have just completed. Cut out two 8¾-inch disks from a 2-inch plank. Screw them together, then turn the resulting disk on a wood lathe until its diameter is 8½ inches. If you don't have access to a lathe, you can do a surprisingly good job with a wood rasp and a sander. Saw out a half-inch strip across the center of the disk and put the two parts back together with a pair of cleats nailed across the gap. When you do this, be sure to allow for the width of the saw cuts. As a safety measure check the disk carefully for absolute roundness. Mount the disk on two standards, using a long bolt placed between the cleats for an axis.

A fiberglass cell (unfinished in this picture) and the pattern from which it was made.

Wrap a 3-inch strip of polyvinyl plastic around the disk, allowing a ½-inch overlap. Tab the ends together with a strip of Scotch Tape. Paint on a fairly thick coat of polyester resin. When this becomes tacky, cover with a 3-inch strip of fiberglass mat, placing the ends flush with one another. Secure it in place with a few turns of nylon fish line. Add a layer of fiberglass cloth and saturate the whole thing with polyester resin. Continue adding layers until you have built the circle up to a thickness of ⅜ inch.

After everything has dried, knock out the mold by removing the cleats. Sand off the edges of the circle and cement it to the base of the cell with resin. Wait for it to dry, then turn the cell over and drill at half a dozen places along the rim. Make everything permanently secure with ¾-inch No. 6 brass wood screws.

Drill the centers of the round projections of the cell and tap these for ¼-inch bolts 1 inch long. Turn the nuts on these bolts all the way down to the bolt heads, then use them for locking devices to keep the bolts positioned after they have been inserted to the proper positions to support the mirror evenly. Place tabs and horizontal-positioning set screws as described earlier for the simple wooden cell.

The end result should be a neat and rugged cell for your mirror, one which will never warp out of shape, corrode, or otherwise deteriorate. The combination of a fiberglass cell in a tube of similar material is an ideal one as far as lightness and strength are concerned, especially if you wish your telescope to be portable.

DOUBLE PLATE MIRROR CELLS

Although many people have used them satisfactorily, the two types of cells described above have a common disadvantage. In each the mirror is supported by three adjustable bolts and held in place by clips. In order to change the orientation of the mirror, these bolts and clips must be loosened and re-adjusted to a new position. Just a slight change in the position of any one of the bolts will shift the cone of light coming from the mirror away from its position on the diagonal mirror. But if you change the position of one bolt, you must change both the others if the mirror is to be supported evenly. The result never seems to place the cone of light exactly where you want it and can lead to a frustrating and time-consuming operation. Why is this true? Because you have possible motion of the mirror in three directions, each at a 60° angle. If you adjust bolt A, the mirror tilts around an axis determined by B; if you move bolt B, the axis becomes the line across C and A; and if you move C, the axis is determined by B and A.

The change in the position of the cone of light can be observed only by looking in the the draw tube (see the section on collimation in chapter 14), and adjusting the bolts requires you to be at the opposite end of the telescope—

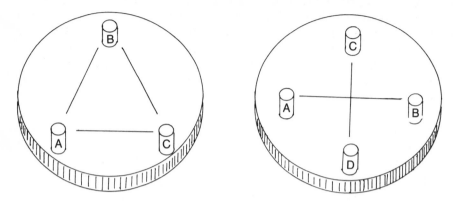

Three-point and four-point support systems for the mirror and/or the base plate of the mirror cell.

but you can't be at both ends simultaneously. If you have a long telescope, the amount of travel back and forth will be considerable. You change the position of one bolt, then move to see what has happened to the cone of light, then move back to change another. Before you have everything lined up satisfactorily, you will have run up a lot of distance on your personal odometer. Of course the process can be done more quickly if you have someone to help, so you can stay at the draw tube while he adjusts bolts.

One way out of this dilemma is to use *four* bolts arranged in a square. If you study the diagram you will see that if you raise bolt A and lower B, the mirror will tilt across the axis between C and D; similarly, raising C and lowering D will create a tilt across axis AB. You will have *two* directions of movement at right angles to each other, making it infinitely easier to achieve the proper orientation of the mirror. But applying four bolts to the back of a mirror can have its own problems, not the least of which is that if you use too much pressure you may damage the mirror. A way to avoid this difficulty is to use *two* plates in your mirror cell, one to hold the mirror in a fixed position and the second to support and adjust the first.

THE CENTRAL SUPPORT CELL

In this type, and in another described below, the top plate is similar to the cell pictured on page 179, but with some important modifications. It can be cut from ½-inch marine plywood and has an aluminum ring attached to its perimeter. Holes are drilled and tapped in the ring to take bolts that prevent lateral movement of the mirror, and clips are added to keep the mirror from falling out when the telescope is tipped too far forward. The mirror is supported by three dowels set into the surface of the plywood and glued in place. Trim the tops of the dowels so the surface of the mirror is parallel to the plate.

Lay out a 6-inch square on the surface of the plate and drill a ⅜-inch hole at each corner. These will accommodate ¼-inch flathead bolts. The bolts must have freedom of motion in the holes, which is the reason for the two different sizes. Bevel the tops of the holes with a shallow countersink whose angle is much greater than that of the bolt heads. Finally, on the bottom of the plate drill a spherical depression just deep enough to accommodate the first quarter inch of a ½-inch steel ball bearing. This may be difficult, but it is very important that a good fit exists between the ball bearing and its socket. The bearing must be at the center of the plate.

Make the bottom or supporting plate from ⅜-inch marine plywood cut just wide enough to fit the inner diameter of the telescope tube. This will eventually be attached to the tube by screws, bolts, or small angle irons. At the center of the top of the plate cut a round depression the same size and depth as the one in the top plate and, using the best epoxy glue you can find, glue the ball bearing into it. Do not use any glue on the top of the bearing. Now drill four ⅜-inch holes to match those in the top plate.

Place the two plates together so they are joined by the bearing and the top pivots freely on the bearing. Insert the four bolts and tighten them in place with lock nuts.

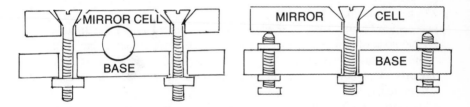

Left, here the mirror cell is supported at its center by the ball bearing and is adjusted in position by four bolts (only two are shown); *Right,* in this form the mirror cell is supported and adjusted by four peripheral bolts (only two are shown) and held firmly in position by the central bolt.

The net result of all this is a unit in which the mirror is held firmly and permanently in position and never need be touched again. But it can be moved in orientation by adjusting the plate on which it rests, simply by tightening or loosening the four bolts as described earlier. It is a unit which will remain solidly in position even if the telescope is jolted or transported from one place to another.

THE FOUR POINT SUPPORT SYSTEM

In the unit described above the mirror rests on a single point, the ball bearing. In another arrangement of the two-plate system the top plate rests on four adjustable bolts and is held in place by a single bolt passing through the centers of both plates. This is again a ¼-inch flathead bolt passing through a ⅜-inch beveled hole and held in place by lock nuts, as shown in the diagram. The four supporting bolts are placed in a square pattern on the bottom plate. In order to make them adjustable, use countersunk hex nuts at both extremities of each bolt with an extra lock nut to keep it in position. Here the bolts, with the exception of the center bolt, fit the holes which accommodate them.

This arrangement, like the central support cell, is very solid and resists jolting and moving. It has the advantage of being much simpler to construct, since it requires fewer bolt holes and doesn't need a hole to fit a ball bearing. And, again like the other cell, it can be adjusted across only two axes instead of three. Which one is used is only a matter of time and preference. Both cells come from suggestions by an expert in the field, Robert E. Cox, and are used here by permission.

THE SPIDER AND DIAGONAL HOLDER

Equal in importance to having a well-constructed primary mirror cell is a rigid, yet adjustable, support for the diagonal mirror. There are many ways to accomplish this and much has been written on the subject. What follows is a method that is a composite of the ideas of several experienced writers.

The Spider

The spider is usually a four-armed structure placed across the upper end of the telescope tube. The arms, or vanes, are wide bands of metal placed edge-on to the incoming rays of light, thus offering little obstruction. This apparatus supports a metal or wood base upon which the support for the diagonal mirror is mounted.

Unfortunately the four-vaned spider produces diffraction patterns around the image. The "star"-shaped projections around the periphery of a bright star image are caused by the slight interference the vanes offer to the passage of light. These effects are much increased if the vanes are straight, because the vane on one side of the diagonal holder reinforces the diffraction of the vane on the other side, and vice versa. One might think that using three vanes instead of four might decrease the number of spikes. Actually it increases it to six, for the reasons mentioned above. A two-vane system would, of course, have only two spikes but is almost impossible to anchor securely. Curved supports produce fewer diffraction effects because they are spread over a larger area of the mirror and thus are less noticeable. Such a system was described in an earlier edition of this book. The curves resulted in an almost diffraction-free image, but it was found that increasing the tension distorted the curves and it was eventually abandoned.

So the four-vane system, in spite of its deficiencies, seems to work out as the best practical solution of the problem. The vanes can be made of brass strips 1 inch wide and 3/32 inch thick. Solder or rivet small right-angle tabs to the end of each vane so it can be attached to the hub at one end and the tube wall at

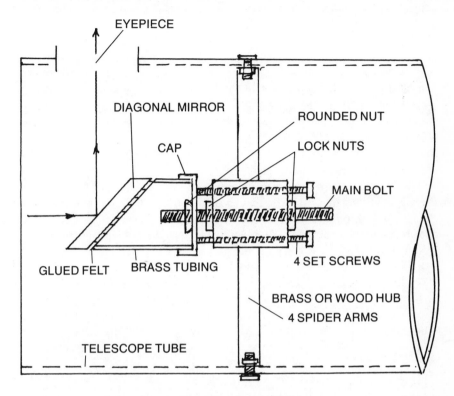

One example of a diagonal holder.

the other. They also must be cut a quarter inch shorter than the distance from the hub to the tube wall. You will also want to displace the hub a short distance from the geometrical center of the telescope tube, for reasons we will explain later.

Make the hub itself from a solid wood cylinder 1 inch in diameter and 1 inch long. It is somewhat better to use a piece of brass of these dimensions, but unless you have a drill and the proper taps wood is easier to work since you can use screws. If you decide to use wood, a section of 1-inch diameter birch dowel is very satisfactory because if you make any mistakes in drilling you simply saw off another piece and start again.

Drill a ¼-inch hole longitudinally through the center of the hub, making sure it emerges exactly at the center of the other end. A 3-inch ¼–20 flathead bolt will later be inserted through the central hole with its head pointing toward the mirror end of the telescope. Using a circular file, enlarge the hole until the bolt can be slid back and forth but still is a fairly tight fit. Now drill four equally spaced holes ⅛ inch from the rim of the hub, and insert ⅛-inch set screws. These should fit very tightly. Any back-and-forth motion should require turning the bolts.

Next, obtain a piece of 1-inch O.D. brass tubing and cut one end square. Cut out a flat piece of brass the same diameter as the tubing; cut a ⅜-inch hole in its center; then, using a metal countersink, bevel the edge of the hole. The bevel must be shallow, in any case less than the angle on the head of the bolt. Attach the plate to the tubing, using solder or small brass 90° angles. Better still, if you can obtain a cap for the tubing, it will make a much neater appearance. Now cut the tubing at an exact 45° angle, starting ½-inch from the cap end of the tube. The diagram will help you sort out these directions.

Finally, cut another piece of brass plate the same size and shape as the diagonal mirror you plan to use. Cut a ½-inch hole in its center. Solder the plate to the 45° end of the tubing.

Now you are ready to assemble the unit. Insert the flathead bolt through the diagonal holder, as shown. Thread on a lock nut followed by a hex nut. Pass the bolt through the hub and thread on another hex and lock-nut combination. Adjust the nuts so the one nearest the diagonal holder is no more than ¼ inch away from it. Then adjust the four set screws until they make contact with the backplate of the diagonal holder. Tighten all nuts and set screws until everything is secure and in line.

The diagonal itself can be attached to its holder in either of two ways. Many telescope makers use three slotted clips screwed to the holder. There are two objections to this practice. One is that the lips of the clamps must extend at least 1/32 inch over the surface of the diagonal mirror. If not, there is the danger of its falling out and damaging the primary mirror. The other is the diffraction they cause in the image. Robert Cox glues his diagonals to the backing plate, using a piece of hard billiard felt as an intermediate layer. The felt is first glued to the backing plate using Pliobond cement, then after it is thoroughly dried, the diagonal is glued to the felt. The result is a firm bond which lasts indefinitely. But this could create difficulties if you wish to clean or re-surface the diagonal. Nevertheless, it is a very good solution to the problem.

Placing the Diagonal Holder

As explained on page 133, the center of the diagonal must be placed 48¼ inches from the surface of the primary mirror. Consequently the spider arms must be secured to the telescope walls at a distance greater than this. How much greater depends upon the dimensions of the unit you have made. It should be at least the distance from the center of the diagonal to the point where the vanes are attached to the hub, plus another ¼ inch to permit fine adjustments.

Once you have decided on the location of the spider arms, drill holes in the telescope tube to take the bolts which hold them in place. Be sure that the holes are placed so that any given arm is a direct-line extension of its opposite vane. Otherwise you will get added diffraction effects. If the telescope tube is thin-walled, wide washers placed under the bolt heads will help prevent distortion, which easily occurs if you tighten the vanes more than you should.

An ingenious use of scrap materials for an 8-inch telescope. (Photo by Michael J. Morrow)

Now comes a rather tricky part of the whole procedure. It was pointed out earlier (page 133) that the diagonal should not be placed in the geometrical center of the tube but should be displaced slightly toward the eyepiece. The mathematics involved in finding the exact distance are somewhat complex, but you can determine it visually. After the eyepiece holder has been placed in position (see next section) look down the draw tube. The edges of the diagonal must appear to be concentric with the edges of the primary mirror. If they are not, you can adjust them by taking up the vanes on the eyepiece side of the tube and loosening the opposite ones. This was provided for in the directions by making the vanes ¼ inch short and compensating for it by using end-bolts.

THE EYEPIECE HOLDER

Once you have decided on the location of the diagonal, cut a hole in the telescope tube wall opposite it. But before you do any cutting, check your measurements again very carefully. Once the hole is cut, you are committed to it. If you get it in the wrong place, it is still possible to atone for the mistake by shifting the positions of the primary mirror and the diagonal, but this is a tedious and tricky operation. To be safe, make the hole ¼ inch wider than the diameter of the eyepiece holder adapter tube.

Cutting a round hole in a curved surface can be difficult. It can be made easier by using a form of the "biscuit cutter" described on page 144. Commercially made hole cutters will also do the job if you can find one of the right diameter.

There are several criteria for a good eyepiece holder. It should fit the surface of the telescope tube snugly (which means its base must be curved to conform to the surface), it should be adjustable to very close tolerances (the range of best focus for an f/7 telescope is only .002 inch), and it should work smoothly and easily. Commercially made holders all satisfy the first requirement, but may vary on the others.

A Simple Eyepiece Holder

The simplest type of eyepiece holder is an adapter tube set in the side of the telescope tube. Inside the adapter is a second close-fitting cylinder called the draw tube. The eyepiece fits into the draw tube and is held in place by friction. Although easy to make and use, this type has two deficiencies which are bothersome. If the adapter is large enough for the draw tube to slide in and out of easily for focusing, it may also permit the eyepiece to fall out when the telescope points the eyepiece downward. The remedy for such a deficiency is obviously a little care in the use of the telescope; yet it is surprising how many eyepieces have been ruined in just this manner. More serious is the fact that it is very difficult to achieve fine focusing with a draw tube. In short, though inexpensive, this type of holder clearly violates both the second and third requirements listed above.

The Threaded Eyepiece Holder

Much better, from all viewpoints, is the threaded eyepiece tube. Here the eyepiece fits snugly into a tube which is threaded into another. Fine focusing is accomplished by turning the threaded tube. An inherent disadvantage of this type is that, if there are many threads to the inch, the difference in the focusing point between nearby and distant objects (which may be several inches) involves a great many turns of the tube. Makers of this type overcome this difficulty by using a draw tube for rough focusing.

Rack-and-Pinion Holder

This type of holder satisfies all three criteria listed above and is probably the best you can obtain. It consists of a base containing an adapter tube which is moved up and down by a rack and pinion device. A draw tube, split longitudinally down one side, fits snugly inside the adapter. Rough focusing is done by moving the draw tube, fine focusing by turning the knob that controls the rack and pinion. However, unless the gear teeth are very close together, a small turn of the knob creates a relatively large movement of the eyepiece. In buying one of these be sure the reverse is true. The greater the the number of teeth per inch the better off you are.

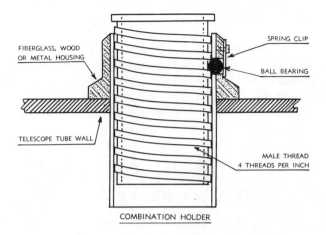

FIBERGLASS, WOOD
OR METAL HOUSING

SPRING CLIP

BALL BEARING

TELESCOPE TUBE WALL

MALE THREAD
4 THREADS PER INCH

COMBINATION HOLDER

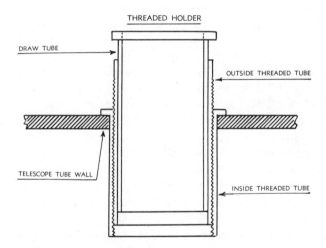

THREADED HOLDER

DRAW TUBE

OUTSIDE THREADED TUBE

TELESCOPE TUBE WALL

INSIDE THREADED TUBE

Two eyepiece holders.

A Combination Holder

This type of eyepiece holder requires a lathe to make, but it combines so many good features that we include it here (see diagram). Its essential feature is a 4-thread-per-inch tube which bears against a ¼-inch spring-loaded ball bearing inserted in a hole in the adapter tube. For rough focusing, the threaded tube is simply pushed in or out, the male threads pushing the ball bearing against the spring so that the passage of the tube is not impeded. Fine focusing is accomplished by turning the tube. The threads now bear against the ball bearing, permitting considerable delicacy of adjustment. This holder is smooth and almost frictionless in its operation.

Making a Threaded Holder

If you decide to make your own holder you will probably need the help of a machinist. The expense involved is compensated for by the fact that the materials needed are cheap and easy to obtain and the cost much less than a comparable model bought commercially.

Our holder will be made up of four parts:

1. A wooden base to hold the adapter tube. The most difficult problem here is the curved base to fit your telescope tube. If you have access to a wood lathe fitted with a cylindrical rotary sander, you can hollow out the bottom of a block 3 inches square by 2 inches deep. This must be done by a trial and error method until you have a good fit. Drill a hole in its center the same size as your adapter tube. Carve the remaining surfaces to your taste.

2. An adapter tube 2 inches long. Ask the machinist to thread the *inside* of the tube, 20 threads to the inch, along its whole length.

3. A second tube to fit inside the adapter. Make this one 3 inches long, and threaded 20 to the inch along two-thirds of its length. These must be on the *outside* of the tube.

4. The third tube fits into the second. Not threaded, it serves as a draw tube and must slide easily up and down. The draw tube must have an inside diameter of 1¼ inches to provide a snug fit for your eyepieces. The diameters of the other two tubes depend, of course, upon the outside diameter of the draw tube. If you have trouble finding three tubes which satisfy these requirements, ask your machinist friend for help.

    Assemble your holder and bolt it to the telescope tube, placing it over the opening so that it is directly over the center of the diagonal. The finished holder should look like the second drawing on page 191.

With the exception of one small modification which will be required to fit the telescope tube into its saddle on the mounting, you have completed work on all its accessories. You need only to build a mounting to hold the tube firmly in any position you wish to place it and all the fundamental work on your telescope will be finished.

CHAPTER 13

The Mounting for the Telescope

I᠇ IS IMPOSSIBLE to discuss all of the variations of amateur-made telescope mountings; but, since each falls into one of several groups, we can examine the characteristics of each group for strong or weak points. All telescope mountings must be constructed to have motion in two directions at right angles to each other. All must be designed so that the telescope can be pointed to any part of the heavens without interference from any supporting structure and must do so without forcing the observer into tiring or awkward positions. Strength, rigidity, and smoothness of operation must be built into the mounting; these must be achieved at the expense, if necessary, of fancy fittings and good looks.

In the first chapter we made brief reference to the two main types of telescope mountings, the altazimuth and the equatorial. We must discuss these more in detail now, for it is important to know the advantages and disadvantages of each.

THE ALTAZIMUTH MOUNTING

This type has the great advantage of a relatively fixed position for the eyepiece, that is, the eyepiece moves up and down with the tube but never becomes displaced sideways, as it does in the equatorial unless special provision is made to compensate for this characteristic. Since the whole altazimuth mounting swivels sideways to take care of the direction (azimuth) of heavenly bodies, and the tube moves up and down in altitude within the mounting to compensate for their height above the horizon (that

193

is, the two axes are vertical and horizontal, respectively), the eyepiece may be placed on the upper surface of the tube or at either side, according to the preference of the maker. This is a great help as far as the comfort of the observer is concerned. This mounting is easy to use, too, because its uncomplicated motion permits it to be shifted from object to object with a minimum of adjustment. Using the altazimuth, you can follow rapidly moving objects such as aircraft or even satellites (if the field of the telescope is wide enough to pick up the satellite in the first place). It is also well adapted for terrestrial purposes if you want to use your telescope for looking around the countryside. The author's first telescope was mounted this way. Located on a New Hampshire hillside, the instrument was used as much to watch the activities at the weather station atop Mt. Washington, 30 miles away, as it was to look at the stars. Finally, this type of mounting is simple and easy to make, needs no counterweights for balance, and is less expensive than the equatorial mounting of comparable strength and rigidity.

But all these advantages are outweighed by a very serious disadvantage. Neither of the motions of the altazimuth matches those of a star or planet as it moves across the sky. The motion of any heavenly object takes place in two directions simultaneously. It is always moving up or down relative to the horizon and at the same time progressing steadily westward (except those in the circumpolar constellations). In order to follow these two motions, the telescope must also be moved in two directions. The result is an oblique motion of the telescope tube if the star is to be kept in the field of the eyepiece. The owner of an altazimuth telescope becomes used to this motion eventually and will make the necessary adjustments automatically, but his visitors (and the possessor of a good telescope has many of these) will have constant trouble keeping a star or planet in the field. Consequently much of the time of an observing session is spent in adjusting the telescope. The final result is that the mechanics of the instrument has interfered with the purpose for which it was built—getting a long and uninterrupted look at the image in the eyepiece.

The double motion characteristic of the altazimuth also makes celestial photography almost impossible. To take a picture of most celestial objects it is necessary to hold the image at one spot on the film for relatively long periods of time. Doing this while moving the telescope up or down and at the same time sideways is in the nature of a physical feat.

The difficulty arises because the apparent motion of the stars is perpendicular to the axis of the earth and therefore varies with the latitude of the observer. When you visualize the earth as a great ball spinning in space surrounded by far-distant stars, you can see why this is true. If your altazimuth mounting were located at the equator its vertical axis would be perpendicular to that of the earth, and at the poles the horizontal axis would satisfy this condition. But at points in between, each axis makes an

Good design in a 6-inch telescope mounting. Made by Howard Snow, a sophomore at Bristol, Tennessee, High School, it is fine evidence that telescope making is not limited to adults. (Photo by Howard Snow)

angle with the line joining the poles. Therefore each axis must be constantly adjusted to follow the motion of the stars.

THE DOBSON TELESCOPE

In spite of the disadvantages of the altazimuth mounting, it is becoming increasingly popular with many amateur telescope makers. This is largely the result of two new developments in the field: the use of plywood as a basic material for mountings, and the advent of large short-focus mirrors which are already ground, polished, figured, aluminized, and coated. Such mirrors are relatively inexpensive. For example, a 12½-inch f/5 costs about $275, and the whole instrument costs less than $400.

The mounting itself is made from simple materials: plywood, tubes adapted from construction forms called Sonotubes, bearings made from pads of Teflon riding on smooth surfaces such as Formica or Marlite for motion in azimuth, and more Teflon supporting large circular discs for altitude bearings. The result is a solid, smooth-working, portable telescope.

This plywood wonder was invented by John Dobson and bears his name. It can be built by anyone reasonably skilled in carpentry. There are only four principal parts to construct: a Sonotube cut to the proper length, a box to enclose the mirror end of the tube, another box to support the tube box and mirror, and a ground board as a base for the instrument. We will consider each in turn.

The Telescope Tube

The Sonotube must have an inside diameter at least 2 inches greater than that of the mirror. Wall thickness of tubes such as this varies with the manufacturer. The thicker the walls, of course, the more rigid the tube. Its length should be an inch or two longer than the focal length of the mirror, but this depends upon where you want your spider, diagonal, and eyepiece to be. The inside should be blackened and the outside covered by at least three coats of varnish or paint, followed by a coat of a polyurethane finish.

The Tube Box

All plywood used in the construction of the wooden parts of the telescope should be ¾ inch or more. This will increase the weight but add to stability. Since all four sides of the tube box must make contact with the tube, its inside cross section will be the same as the outside diameter of the tube. This is very important, because the tube shouldn't rattle around in its supporting box. Cut the sides of the box the exact diameter of the tube, and the top and bottom boards 1½ inches wider. Here "top" means the board on top of the tube when it is horizontal (see diagram). The front board will be square, the same dimension as the outside width of the box. Cut a hole in the front board the same diameter as the tube. Take care here, for this board must fit exactly. The box must be long enough to accommodate the bearing, which in most cases should be 6-inch diameter PVC sections attached near the top of the box. Most builders use PVC, but any circular structure may be used as long as it has a smooth, hard, unyielding rim and is absolutely uniform in diameter. Don't attach the bearings at this point; you won't know exactly where to place them until you know the exact balance point of the telescope tube with its box, mirror, eyepiece holder, spider and diagonal, and finder scope. Finally, attach corner braces inside the four corners of the bottom of the box. The inside surfaces of the braces must be curved to conform to the tube. When all this is done, reach inside the tube and nail it to the box so there is no possibility of the tube sliding down when the telescope is in use.

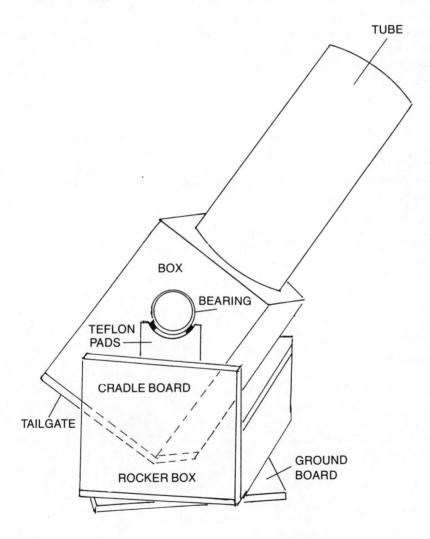

The basic form of the Dobson telescope.

The Cradle Boards

These boards, one on each side, support the bearings and therefore the telescope tube. They are attached to the inside of the walls of what is called the rocker box. Each board will have a semicircular arc cut at its top whose diameter is the same as that of the bearings. Actually, these should be less than a semicircle—what the geometers call a circular arc, as shown in the diagram. On the inner sides of the arcs, near the top, nail two Teflon pads ¾ inch wide and 3 inches long. Use finishing nails, making sure their heads are sunk below the surface of the pads. The width and length of the cradle boards is a matter of choice, but they must be long enough to allow plenty of clearance for the telescope tube and its box.

The Rocker Box

The name of this box is perhaps not well chosen because it doesn't rock. Instead it rests solidly and securely on the Teflon pads of the ground board, and its only motion is circular.

The dimensions of the rocker box, especially its width, are critical to the smooth operation of the telescope. The side boards and the bottom should be 1 inch thick. A front board is also 1 inch but does not extend all the way to the top of the box. This is to permit the telescope to be tipped to a horizontal position. The height of the box may be as much as 18 inches, but this is not a critical measurement.

Now for the width. It must be the total of the outside width of the tube box, *plus* the sum of the bearings, *plus* the sum of the thickness of the cradle boards, *plus* an allowance for clearance, about ¾ inch. The clearance is included so the tube box will not rub against the cradle boards. The total is the *inside* measurement of the rocker box; its outside dimension must be 2 inches wider since the sides are 1 inch thick. The length of the rocker is arbitrary, but it should be several inches longer than the diameter of the telescope tube. The bottom board will have the same dimensions as those of the outside of the rocker box.

It is essential that the rocker box be very strong because it is subjected to a great amount of stress. In general, whenever two pieces of wood are joined, use white carpenter's glue, finishing nails every three inches, and clamps or weights to hold them together until the glue is completely dry.

The final addition to the rocker box is a piece of Marlite, a hard, waterproof material often used as shower walls. Or you can use Formica, easily obtained from dealers in kitchen counters. Formica sink cut-outs are ideal for this purpose. Cut out a section the same size as the bottom board and glue and nail it to the bottom of the rocker box, smooth side down. Don't use any nails within a circle the same size as the telescope tube, using the center of the bottom board for the center of the circle.

A variation of the Dobson telescope. (Photo by T. J. Britton)

The Ground Board

Make this of two pieces of 1-inch plywood, glued and nailed together. It should be the same width and length as the bottom of the rocker box. Cut three pieces of 1-inch plywood, each 2 inches square. Nail one at each corner of the long end of the board, and one at the center of the opposite end. The three now form a triangle. Find the center of the triangle and drill a ⅜ inch hole. Drill another ⅜ inch hole at the center of the bottom of the rocker box. Later a bolt will hold the rocker box firmly to the ground board.

Turn the board over and, using the bolt hole as a center, draw a circle the same size as the diameter of the telescope tube. At 120° intervals, nail three 2-inch-square Teflon pads to the rim of the circle. Again, make sure the nail heads are below the surface of the Teflon. These pads act as bearing for the smooth surface of the bottom of the rocker box. Insert the bolt, threaded end up, through the two matching holes. Use a washer and a locking nut to be sure the bolt won't become loose. The rocker box should now rotate smoothly and freely on its Teflon bearings.

The Mirror Cell

Since many Dobson builders like to use large mirrors (10 to 16 inches), great care must be taken to see that they are well supported. Commercial mirror makers often use thin blanks from which to make their products. These need 9-point suspension rather than the three used for thicker ones. Further, in equatorial mountings the weight of the mirror shifts from side to side as the telescope is turned in various directions, but in the Dobson weight is always directed toward the underside area of the tube. If the mirror is thin, this can cause a distortion of the image if only one side support is holding the mirror in position. To avoid such a condition, Dobson mirrors can be supported by passing a strap around the whole lower hemisphere, thus distributing the weight evenly.

Otherwise, the mirror cell can be made with a double-plate support as described earlier, the top plate (with the additions mentioned above) holding the mirror firmly in position, and the bottom plate attached to the end of the tube box. In this case, of course, the bottom plate will be square, its edges flush with the sides of the box. One suggestion is to attach the bottom plate as a tailgate, hinged on the lower edge, and to use clamps to hold the other three sides firmly in position.

At this point it is a good idea to attach the other fittings to the tube: finder, spider and diagonal, eyepiece holder, etc. Then, using a sling, find the exact balance point of the combination. Now you know the points where you will place the centers of the circular bearings.

The final step is the placement of the cradle boards. A line passed through the low point of arcs from one side to the other must pass directly over the bolt holding the rocker box and ground board together. The cradle boards must also be high enough so that the mirror end of the tube has a clearance of at least 2 inches when the tube is vertical and does not rest on the front of the rocker box when it is horizontal. Use screws to attach the cradle boards to the inside of the rocker box.

Place the telescope in position and check that the PVC circles work smoothly on their Teflon bearings. If the movement is too stiff, remove the telescope and move the pads toward the bottom of the arcs. Check also to be sure that the whole unit rotates smoothly on its ground board bearings.

THE EQUATORIAL MOUNTING

If you could construct a mounting in which one axis is parallel to the earth's, a point on this axis would sweep across the sky in the same direction as the stellar motion. If you place your telescope tube on this point, it also would follow a star as the axis turns. This is the principle of the equatorial mounting. The main axis is fixed parallel to the line joining the poles, and is called the *polar axis* for this reason. The other axis, called the *declination axis,* is the one on which the telescope tube is mounted. It in turn is mounted on the polar axis at right angles to that axis.

Let us take a moment to explain declination a little further. We can think of the stars as placed on the inside of a great hollow globe called the celestial sphere. Now imagine a plane cutting the earth at the equator and extending into space. Eventually this plane will cut the celestial sphere along a line opposite our equator, called the celestial equator or equinoctial. In other words, the celestial sphere is a large copy of our own except that it has names which differ for purposes of identification. Thus we speak of the north and south celestial poles and the equinoctial, and we measure distances between these points in degrees of declination instead of degrees of latitude.

If the polar axis of the telescope is fixed parallel to the earth's axis, it is of course parallel to that of the celestial sphere. Placing the telescope tube perpendicular to the polar axis will cause it to point toward the celestial equator, or 0° declination, while if the tube is placed parallel to the polar axis (the earth's or the telescope's) it will point toward either the north celestial pole (90° north declination) or the south celestial pole (90° south declination).

Now if we mount the tube on an axis which is at right angles to the polar axis, we can equip it with a scale so that we can point the tube to any desired declination point in the sky. The declination of any star is constant for all practical purposes; thus if the tube is adjusted on its declination axis to point toward a star, we need only to turn the telescope around its polar axis to follow the star across the sky. In short, an equatorial mounting makes it possible to keep a star in the field by moving the telescope in only *one* direction instead of *two*.

This is obviously a great advantage, for if we attach a slow-motion device to the polar axis to turn it at the same rate as the stars (approximately 1 degree every 4 minutes), any given star will stay in the field indefinitely. This is a great boon for the telescope owner, for he can center an object like the Hercules Cluster in his eyepiece, start his slow-motion drive, and stand back while a long procession of visitors look to their heart's content. Or he can attach a camera instead of a visitor to the eyepiece, secure in the knowledge that the light from the celestial object is building up an image on the emulsion of his film.

These are the reasons why most telescope makers prefer to build an equatorial instead of an altazimuth mounting in spite of the fact that the former is a little harder to construct accurately. The first question to be answered is how to determine the angle at which the polar axis must be placed so it will be parallel to that of the earth. The answer is simple: it is equal to the latitude of the observer. You can demonstrate this for yourself by drawing a circle and measuring the angles with a protractor. If your latitude is 30° N, a line drawn from your location to the center of the earth makes an angle of 30° with the plane of the equator. The plane of your horizon—that is, a plane represented by

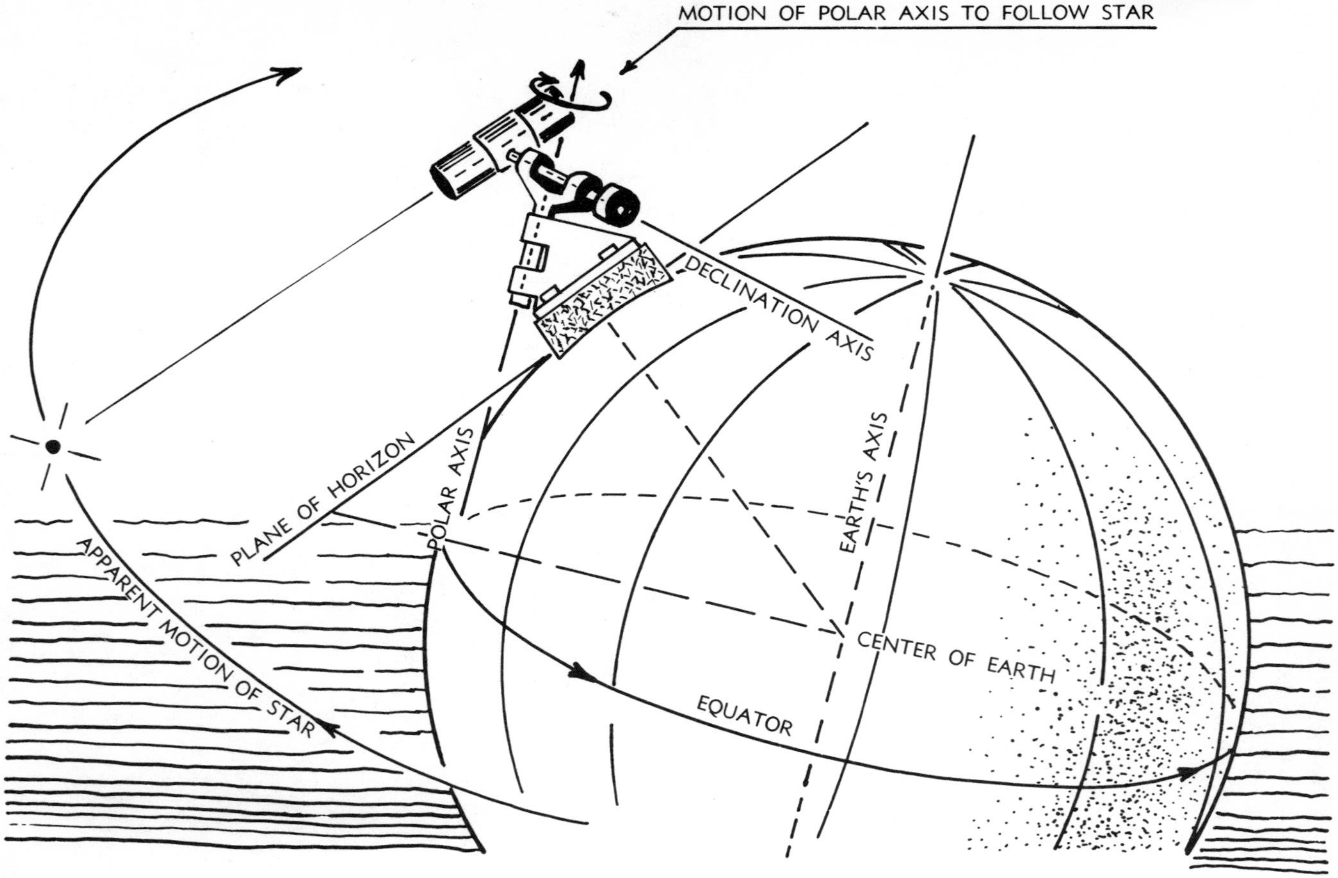

The principle of the equatorial mounting. (See also pages 195 and 227.)

MOTION OF POLAR AXIS TO FOLLOW STAR

DECLINATION AXIS

EARTH'S AXIS

CENTER OF EARTH

EQUATOR

POLAR AXIS

PLANE OF HORIZON

APPARENT MOTION OF STAR

the horizontal line produced by a carpenter's level—is perpendicular to this line. Application of simple geometry to the right triangles involved in the diagram shows that a line parallel to the axis of the earth must also make an angle of 30° with the plane of your horizon.

You can find your approximate latitude by measuring the altitude of the North Star above the plane of the horizon, which will give you a figure within one degree. Easier and more accurate than this, though, is to find your location on a U.S. Coast and Geodetic Survey map which gives your latitude within one minute of arc.

THE POLAR AXIS

The fundamental requirement of the equatorial mounting, then, is that the polar axis be placed at an angle equal to the latitude of the observer. Once this is done, and provision made for it to rotate in its bearings, the polar axis can be made in many different forms. It can consist of a yoke in which the tube is suspended, a two-pronged fork with the tube placed between the prongs, or a single shaft crossed by another in the form of a "T," in which one end of the crossbar of the "T" supports the tube and the other is counterweighted for balance. Each of these forms has its own virtues.

The Yoke Mounting

This is perhaps the easiest mounting to construct because it requires no counterbalancing other than that needed to balance the tube itself.

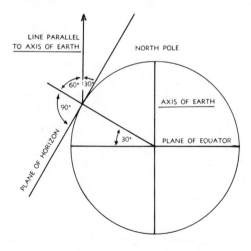

The polar axis points toward the North Star if its angle with the horizon is equal to the observer's latitude.

Its essential element is a yoke supported at each end, with the tube mounted to move in declination between the sides of the yoke. It needs only four bearing points, one on each end of the yoke and one on each side of the tube. It is strong and rigid because of the wide separation of the bearing points, especially those on the polar axis. But it also has several deficiencies. First, the tube cannot be rotated radially in the yoke, and this results in some very awkward eyepiece positions for the observer. The difficulty can be overcome, of course, by providing a special ring in which the tube can rotate, or by providing a rotating ring on which the eyepiece and spider can be mounted. But each of these alternatives calls for special construction work. Second and most important, the yoke mounting makes it impossible to view constellations near the polar axis. Finally, this type of mounting is usually bulky and heavy with a resultant lack of portability unless the maker also happens to own a truck. Many amateurs are willing to overlook these difficulties (or fail to take them into account) because of the simplicity of construction.

The Fork, or English, Mounting

This, like the yoke mounting, needs no counterbalancing except in the tube and consequently is very popular with many amateurs. The tube is suspended in declination between the open prongs of the "fork," whose shaft is supported by bearings. To reach the polar regions with this type of mount requires (1) that the prongs of the fork be long enough for the mirror end of the tube to swing between them without hitting their junction, or (2) that the suspension points for the tube be placed very close to the mirror end, in which case extra weights must be placed near the end of the tube to achieve balance. Each of these conditions is likely to introduce vibration in the tube.

A third possibility is an offset fork with the tube placed far enough from the shaft of the fork so that the telescope can be turned parallel to the polar axis. When this is done, however, a counterbalance is required and the shaft must extend far enough from its bearings to prevent interference with the tube motion. This is likely to introduce vibration in the shaft itself, although with a light fiberglass tube it can be prevented if the shaft is heavy enough.

The German Mounting

This is the most popular mounting among amateurs in spite of the fact that it requires counterbalancing. There are many variations in the basic design, but in essence it consists of two shafts arranged in a "T" with the telescope tube mounted to rotate in declination on one end of the cross bar of the "T" while the other end is counterweighted. The polar axis, or vertical part of the "T," rotates in bearings mounted on a slanting base. The angle of the base is equal to the latitude of the observer.

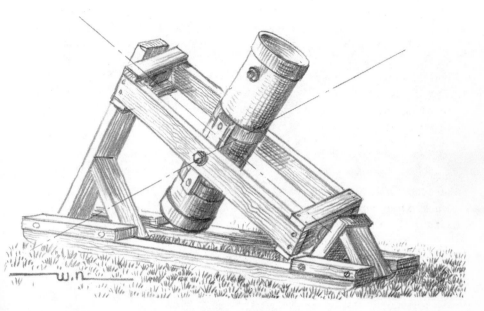

The yoke mounting.

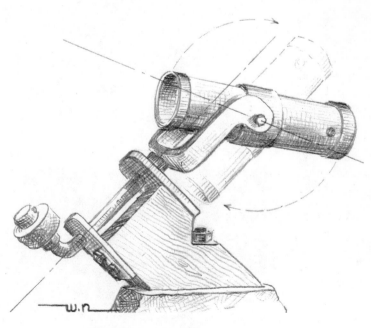

Offset fork mounting.

Portable mounting for an 8-inch reflector, made by John Mierzwa, of Auburn, New York. This mounting is very strong but so light that it can be carried by Mr. Mierzwa's eight-year-old daughter. The drive has a friction clutch so that the telescope may be adjusted even while the drive is running. (Tarby Photo, Auburn, N. Y.)

Fork Mounting for a fine 12-inch f/4.3 telescope, made by George T. Keene, Rochester, New York. This is the telescope used for the excellent photographs shown in chapter 17. Note the use of 3-inch pipe fittings for the fork. The drive is shown in the photo below. (Photo by George T. Keene)

The German mounting has many advantages because of the simplicity of its design. It can be made from ordinary pipe fittings, which makes possible a rugged and stable mounting at very low cost. Because the shafts are easily accessible, they can be fitted with accessories such as setting circles and slow-motion drives. The base upon which the polar axis rests is compact and can be mounted on either a tripod or a permanent pedestal. Its sole disadvantage is that its motion is somewhat clumsy and awkward to handle because the tube is sometimes on one side of the mounting and sometimes on the other. But this is of minor importance compared to its many virtues, and it is the type recommended here. We describe two variations: one that is perhaps the simplest and least expensive mounting possible to make and still retain ruggedness of construction, the other a more elaborate form of the same design.

MAKING A SIMPLE GERMAN TYPE OF PIPE MOUNTING

Materials needed:

1 $2\frac{1}{2} \times 12$-inch pipe, threaded at each end (all pipe in this list is galvanized)
2 pipe caps for $2\frac{1}{2}$-inch pipe
2 pipe-mounting brackets
1 union
1 close nipple
1 tee
1 $2\frac{1}{2} \times 18$-inch pipe, threaded at each end
1 $2\frac{1}{2} \times 4$-inch pipe, threaded at each end
1 pipe flange
1 piece $\frac{3}{4} \times 12 \times 18$-inch marine plywood
2 pieces $\frac{3}{4} \times 6 \times 12$-inch marine plywood
1 piece $4 \times 6 \times 6$-inch No. 1 hard pine

The Base

Cut the piece of hard pine along its 6×6-inch side to produce an angle equal to your latitude. You can do this using a protractor, in which case you can probably measure within a degree of your latitude or, to be more accurate, use simple trigonometry to make measurements for the cut.

In any right triangle, the relationship between the base and the height can be expressed as the tangent of the angle opposite the height, or:

$$\text{tangent of angle } X = \frac{\text{height}}{\text{base}}$$

To use an example, suppose your latitude is 41° 51′. From a trigonometry table you find that the tangent of this angle is .8957. Since

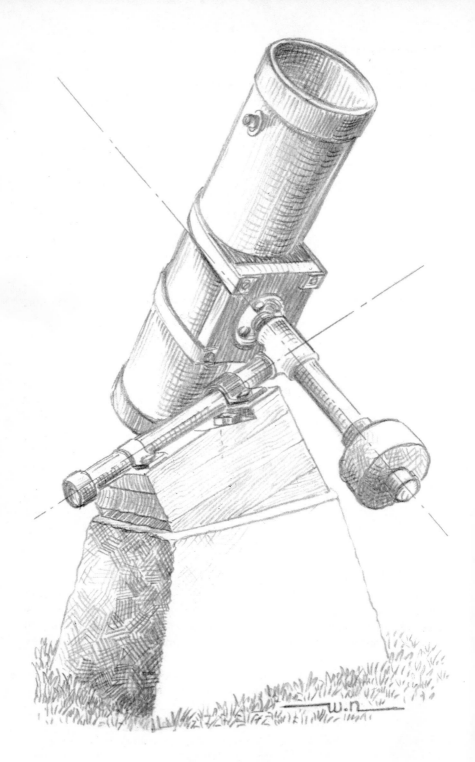

German mounting.

the base of the block will also be the base of the triangular support, it will be 6 inches. When we substitute these quantities in the formula, we obtain

$$.8957 = \frac{\text{height}}{6}$$

hence

$$\text{height} = 5.3742 \text{ inches} = 5\frac{3}{8} \text{ inches}$$

Measure this distance carefully along one end of the block (be sure the block is square first, that is, that all its angles are right angles), draw lines, and saw it. Sandpaper all surfaces, being careful that you don't change the angle by too vigorous application of the sandpaper. Seal the pores of the wood with two applications of shellac, then paint or varnish as you see fit.

Attach the $2\frac{1}{2} \times 12$-inch galvanized pipe to the sloping surface of the block with the pipe brackets, using bolts rather than screws to be sure that the pipe is held very firmly against the wood. Allow an equal overhang of the pipe at each end of the block. Screw the pipe cap on the lower end of the pipe.

At the upper end, screw on the pipe union and inside this, the close nipple. Now apply a mixture of No. 220 Carborundum and oil to the threads of the nipple and the shaft of the tee. Work the threads against each other until all minor burrs and rough spots have been worn away and the threads fit with no play or tendency to grip or bind. Wash away all traces of the oil-grit mixture and grease the threads, using a light grease. The turning of the tee on its supporting pipe is the polar axis of your telescope.

Screw the 18-inch pipe into one side of the tee, closing the other end of the pipe with a cap. Then screw the 4-inch section into the other side of the tee. Using more oil-grit mixture, make sure that the threads at the end of the short pipe section fit those of the pipe flange.

The Saddle

Cut semicircles in the two pieces of $\frac{3}{4} \times 6 \times 12$-inch plywood, making the diameter equal to that of the tube. Mount the two pieces at opposite ends of the $\frac{3}{4} \times 12 \times 18$ and brace them firmly, using angle irons or triangular braces cut from the plywood. The saddle for the tube, made in this way, must be very rigid, so use glue and the longest possible screws (at least $2\frac{1}{2}$ inch at all main points of attachment) for this. A piece of steel tape, long enough to pass over the tube when the tube is attached to the arms of the saddle, serves to hold the tube firmly in its saddle. The mainspring of an old eight-day clock is a good source of steel tape, another source, although not as pleasing to the eye, is ordinary pipe strap. Drag an old trunk out of the attic and remove one of its snap latches. Rivet one end of the steel tape to the short side of

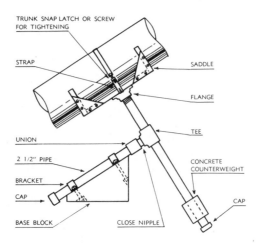

Schematic drawing for a simple pipe mounting.

the latch. Screw the longer side of the latch to the center of one side
of the saddle. Attach the other end of the tape to the opposite side
of the saddle after tightening it to a point where it forms a snug fit on
the tube when the latch is closed. This fit should be just tight enough
so that when one end of the tube is grasped in both hands and a rotary
force is applied, the tube can be turned to any desired position in the
saddle. As we have pointed out before, permitting rotation of the tube
in the saddle is a very necessary requirement of a good mounting; otherwise the eyepiece assumes very awkward positions for the observer when
the tube is pointed at certain sections of the sky.

Make and attach the finder (see page 223) and find the balance point
of the tube as described on page 225. Bolt the pipe flange to the center
of the saddle, then screw the whole unit on the end of the short pipe
on the mounting. Because this will serve as the declination axis, you must
leave room on the threads for the saddle to turn freely. Check the saddle
for longitudinal balance at all possible positions of the axis. If necessary
add small counterweights to the underside of the saddle until you can
make it balance in all positions. Now attach the tube with all its fittings
to the saddle by means of the steel tape and latch so that the center of
gravity is lined up with the axis of the declination shaft.

Place the declination axis in a horizontal position and tie an empty
sugar bag to a point 4 or 5 inches from the end of the long pipe of the
mounting. Fill the bag with sand until it just balances the tube. Mark
the position found; then turn the axis through 180° and balance it again.
The weights should be the same amount and fall in the same position
for both positions of the declination axis. If they do not, take the average
position and the average weight for the construction of the counterweight.

The Counterweight

If your tube and saddle are not too heavy, you can make a very serviceable counterweight from concrete. Place a short piece of well-greased 2½-inch pipe in the center of a No. 10 tin can. Weigh them carefully on the bathroom scale. Now fill the can around the pipe with a concrete mix until the total weight, minus the weight of pipe and can, just equals that of the bag of sand you used for a trial counterweight. When the concrete hardens, remove the pipe and strip off the can. The resulting lump of concrete when smoothed, dried, and painted, will make a counterweight which will fit on the long end of the declination axis. You can fix it on its balance point, as you add more gadgets to the telescope tube, by drilling a series of holes along the pipe on which it rests. If properly spaced, pins or bolts through any two of these holes will keep the counterweight firmly in place.

If properly made, a concrete counterweight is a good-looking and efficient device. Some workers object to its bulkiness, which may assume large proportions if the tube and fittings which it balances are heavy. In this case, it is probably better to use lead or steel. The sawed end of a steel billet makes an excellent counterweight, but you are faced with the problem of drilling out a lengthwise hole large enough to accommodate the declination shaft. This is a formidable problem if the shaft is large, as in the case of this simple mounting. On the other hand, once drilled, the steel counterweight is easily held in place by drilling and tapping a hole in its side for a ⅜-inch set screw which bears against the shaft.

Simpler to make, but more expensive, is a lead counterweight. You can cast it yourself by borrowing a melting pot and blowtorch from a plumber and using a tin can as a casting vessel. A birch dowel the size of your shaft can be nailed upright in the bottom of a tin can and will provide the necessary hole in the casting. Weigh out the necessary number of lead ingots—these can be obtained from a plumber or ordered through a hardware store—melt them in the pot at the lowest possible temperature, and pour the molten lead around the birch dowel. When the casting has cooled enough to handle, knock out the dowel and strip off the tin can with a pair of heavy pliers. Lead can be drilled and tapped like steel but with much more difficulty since the lead fouls a drill very quickly. If you are careful, though, you can drill a hole in the side of the casting to hold it in place on the shaft with a set screw.

The Mounting Pedestal

Now comes the decision as to what to use for a base for your mounting. If you want the telescope to be portable you can build an inexpensive tripod, or buy one from war surplus stock if you can find one heavy enough for the purpose you have in mind. Having gone to considerable trouble to make a rigid, vibration-free mounting, you will not of course wish to waste all your effort by placing it on a flimsy tripod. So if you decide to buy a tripod, be sure its primary characteristic is strength. Then make sure it is high enough that when the telescope is turned toward the zenith of the heavens, the end of the telescope will not hit the ground or strike the legs.

A Tripod

Making your own tripod is not a difficult task. You need an 8-inch circle of 2-inch oak in which you will cut "ears" as points of attachment for the tripod legs as shown in the diagram. Through each ear drill a hole for a $\frac{1}{4}$-inch carriage bolt. The legs can be made of 1×3-inch No. 1 pine stock. Cut six pieces to length, at least 6 inches longer than the distance from the center of the point of attachment of the tube in its saddle to the end of the cell. Arrange the pieces in pairs and bolt the members of each pair together 2 inches from one end, using $\frac{1}{4}$-inch round-head carriage bolts. Drill $\frac{1}{4}$-inch holes $\frac{1}{2}$ inch from the other end of each pair. Spread these ends apart and attach them on opposite sides of each ear of the tripod head. Round off the upper ends of the legs to prevent them from binding against the base of the ears. Use round-head carriage bolts here, too, with lock washers and wing nuts so they can be tightened by hand. At the middle point of each leg place spreaders between the leg sections to prevent warping or distortion. Use a hardware store 1-inch ring as a centerpoint for three pieces of window-sash chain to limit the spread of the legs. Be sure the legs are far enough apart to provide a stable support for the mounting, but not so far that they interfere with the action of the tube as it is moved to various positions. To insure stability, a 60° angle with the ground at the base of each leg is usually sufficient.

Drill the angled base block of the mounting on each side of the polar axis for a tight fit for $\frac{1}{4}$-inch bolts. Bore out corresponding holes in the head of the tripod. Use carriage bolts, their heads countersunk in the base block and their ends projecting through the tripod head. Again use wing nuts held in place with lock washers so that the mounting can be detached easily from the tripod for transportation purposes. The carriage bolts should be a permanent part of the base block and therefore should be glued in place so that they will not slip out of position when the mounting is placed on or removed from the tripod.

A tripod made in this way is not adjustable for height or for uneven terrain, but there is little need for these features with a telescope. If the ground on which the telescope is placed is so rough that it requires an adjustable tripod, it will be a poor place for an observer moving around in the dark. Adding adjustable leg sections is not a very difficult task but it is generally unwise to do so since additions of this kind usually produce instability in the tripod.

In spite of the fact that a properly constructed tripod offers all the support necessary for the mounting, the use of one is not recommended unless it is necessary to transport the telescope some distance to obtain good seeing conditions. Moving a Newtonian any more than is necessary is never a good idea: the optical parts may get out of alignment and the size of the instrument makes for difficulty in transportation and usually results in rough treatment of the instrument. Another difficulty is that the polar axis must be lined up on the North Star in each new location if the telescope is to work properly in right ascension. This in turn requires a level location for the tripod. All these drawbacks may be avoided, of course, if you are lucky enough to have a reasonably good observing site in your back yard. In this case you need only to set a 4-inch pipe in a concrete base—get it below frost line if you live in a cold climate—and put a flat cap on it. Attach the base of the mounting to the cap in its permanent position. When the telescope is not in use you can store the tube indoors after removing it from the saddle. Cover the mounting with a polyethylene bag when it is not in use.

Before all this can be done, however, you must assemble the instrument. Place the mounting on the tripod head or post and fasten it in place with the wing nuts and lock washers. Place the counterweight on

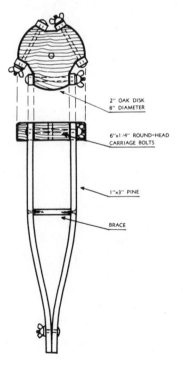

2" OAK DISK
8" DIAMETER

6"x1/4" ROUND-HEAD
CARRIAGE BOLTS

1"x3" PINE

BRACE

Tripod: head and leg detail.

the declination arm and allow it to swing down to its lowest position. Now place the telescope with all its attachments—mirror and cell, spider and diagonal, eyepiece holder, finder, etc.—in its previously determined position in the saddle of the mounting. Snap the restraining latch tight. Move the counterweight to the position where it balances the tube as it swings around the polar axis. Fix the counterweight in this position. If the optical parts have already been lined up (see chapter 14) your telescope is all ready to use.

MAKING A BALL-BEARING GERMAN MOUNTING

The mounting described in the previous section is satisfactory in all respects except that its control depends upon the friction of the threaded pipe fittings against each other which, in turn, requires excellent balancing of tube and supporting structures. Besides this, it is difficult to fit it with setting circles or slow-motion devices.

Another design, which is a little more expensive to build but which can be controlled more easily and can also be adapted to the features mentioned above, may also be constructed from pipe fittings. The moving parts work on ball bearings; there is a friction clutch on the polar axis and a locking knob on the declination axis. The ball-bearing feature makes the instrument so easy to control that the friction clutch and locking knob become necessary additions to the design. Some workers prefer babbitted bearings for their telescopes but find it difficult to line up axes of the telescope while the bearings are poured. If self-aligning ball bearings are used instead, this difficulty is avoided and the construction details are simpler and very little more expensive. The main features of the mounting are shown in the diagram.

Materials needed:

2 2-inch close nipples
1 tee, 2-inch openings at ends, $1\frac{1}{4}$ inch at side
2 2-inch pipe flanges, 6 inches wide
1 $1\frac{1}{4}$-inch pipe flange
2 flange bearings, self-aligning, for $1\frac{5}{8}$-inch shaft
1 4-foot length of $1\frac{5}{8}$-inch steel shafting
1 worm gear, with matching worm, shaft, and bearings
3 1-inch circles of $\frac{3}{8}$-inch brass or plastic
1 4-inch circle of $\frac{3}{8}$-inch steel plate
2 $1\frac{5}{8}$-inch collars for shafting
2 pillow blocks, self-aligning, for $1\frac{5}{8}$-inch shaft oak or plywood as indicated in directions below

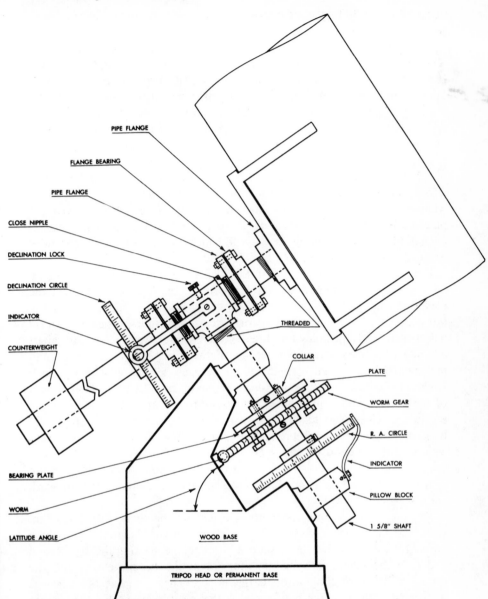

PIPE FLANGE

FLANGE BEARING

PIPE FLANGE

CLOSE NIPPLE

DECLINATION LOCK

DECLINATION CIRCLE

INDICATOR

COUNTERWEIGHT

THREADED

COLLAR

PLATE

WORM GEAR

R. A. CIRCLE

INDICATOR

PILLOW BLOCK

1 5/8" SHAFT

BEARING PLATE

WORM

LATITUDE ANGLE

WOOD BASE

TRIPOD HEAD OR PERMANENT BASE

Design for an equatorial mounting for telescopes up to 12 inches in diameter.

The Polar Axis

The flat-surfaced base block described for the simple mounting in the previous section must be replaced by an indented base section to allow space for the worm gear. This section can be cut from a single block of wood or built to the proper dimensions (see diagram). The indentation between the projecting supports for the pillow blocks which hold the shafting must be deep enough to permit space for the worm gear and worm. It must also be cut at the same angle as that between the two projections, or equal to the latitude. All surfaces should be well sealed with shellac and varnish to eliminate any trouble from moisture.

Cut off a 12-inch piece of the 1⅝-inch round stock for a polar axis. Ask a plumber to cut pipe threads on one end of this piece and also on one end of the remaining section. Mount the pillow blocks on the projections of the base block with bolts or heavy lag screws. These pillow blocks are made so that their contained bearings can be tipped slightly in any direction to permit lining up the shafts which they hold. The bearing part of the pillow block has a collar in which set screws are placed to hold the shaft securely in place, but to insure that the shaft cannot shift in the bearing it is a good idea to drill the shaft just deep enough for the set screws to get a toe hold.

Bolt a collar to the center of the 4-inch steel circular plate. If you have a machinist friend, ask him to drill a 1⅝-inch hole through the center of the plate and also to face off the side of the plate on the other side of the collar. In the worm gear (sometimes called the worm wheel) drill and tap holes for three ⅜-inch set screws placed 120° apart at points 1½ inches from the center of the gear. Insert set screws and nuts in these holes and cap the point of each set screw with a 1-inch circle of ⅜-inch brass. Your machine shop friend can cut these out of brass stock and drill a hole part way through to take the end of the set screw. These should be at least ⅜-inch set screws, since they will have considerable lateral pressure on them. The brass plates will bear against the 4-inch disk and will work very smoothly after they have seated themselves. If brass seems like an extravagance here, you can use 1-inch circles of Lucite or similar plastic. We have found this to work very well.

Slip the steel disk and collar arrangement on the shaft, then the worm gear, and finally the second collar, as shown in the diagram. With these parts still loose on the shaft, slide it through the bearings, first one end and then the other, with the smooth end at the lower bearing and the threaded end projecting through the upper. The fact that the bearings move within their pillow blocks makes this easy to do. Adjust the worm gear to a central position on the shaft, then tighten the set screws of the shaft collars until they are very strongly held in place. Allow enough clearance on the upper collar so that the set screws in the worm gear can bear against the brass or Lucite plates, which in turn exert pressure on the 4-inch disk. With this arrangement the worm gear, through the friction of the brass plates against the 4-inch disk, can turn the shaft. The opposite is not true, however, for the shaft may be turned independ-

ently of the worm gear since enough pressure will cause the surface of
the disk to slide over the brass plates. The amount of friction between
brass plates and steel disk may be varied by adjusting the set screws
which hold these elements against each other. The final effect of this arrangement is that the telescope, mounted on the declination axis which
is in turn connected to the end of the polar axis, will turn in slow motion
with the worm gear but is also free to turn independently by manual pressure applied to it.

The worm shaft is mounted in bearings attached to the base of the
mounting. You must be careful here since too much play between the
worm and the worm gear will result in backlash. At the same time too tight
a fit makes the worm hard to turn. You can make sure of good adjustment
by using shims under the bases of the bearings so you have exactly the
right amount of engagement between the teeth.

There are several advantages in this kind of arrangement for the polar
axis. It is relatively easy to attach a cable to the end of the worm shaft
so that you can turn the telescope in slow-motion right ascension from
your position at the eyepiece. And a gear can be attached on the worm
shaft to drive the telescope by means of a motor or other driving mechanism. If you attach a motor, the knob on the slow-motion shaft can be used
to speed up or retard the drive manually for tracking purposes. Setting
circles are clearly visible because of the open construction; if mounted
at any point on the polar shaft and if the graduations are large enough,
they can be read from the observer's position. Finally, the mounting is
sturdy and solid and is exceptionally free from vibration.

The Declination Axis

Screw the side opening of the tee to the projecting threaded end of
the polar shaft. In case you wonder how a $1\frac{5}{8}$-inch shaft can be screwed
into the $1\frac{1}{4}$-inch opening at the side of the tee, the explanation lies in
the peculiar way in which plumbers measure their pipe fittings. A $1\frac{1}{4}$-
inch pipe actually has an outside diameter of 1.660 inches, and it is the
outside measurement in which we are interested. The $1\frac{1}{4}$-inch opening in
the side of the tee, therefore, is actually a little *larger* than the size of the
$1\frac{5}{8}$-inch shafting, but this need be no cause for worry since the difference is small and the threads on the shaft will still fit securely in the tee
opening. Use a pipe wrench to make sure this fitting is tight. Insert the
close nipples in each end of the tee.

Bolt a pipe flange on each of the two flange bearings. It will be necessary to drill holes in the bases of the pipe flanges for this, and you should
drill them a little larger than the size of the bolt you intend to use so that
you can make lateral adjustments later to line up the shaft of the declination axis. Now screw the pipe flanges to the nipples projecting from the
ends of the tee.

Slide the 1⅝-inch round stock through the bearings until the threaded end projects 1½ inches from one of the bearings. Remove the set screws from the bearing collars and mark the position of each on the shaft with a prick punch. Drill the shaft at these points with holes whose depth and width correspond to the size of the set screws. Replace the shaft and make sure the set screws are seated in their holes.

Check the position of the shaft to be sure that it is at right angles to the polar axis. This is difficult to measure since a square cannot be used because the pipe fittings get in the way. But you can check it accurately simply by rotating the declination shaft around the polar axis, making sure that the shaft has no tendency to wobble at different points in the circle it describes. If a wobble is evident, you can remove it by shifting the positions of the flange bearings on their pipe-flange bases. Once you are satisfied that the axes are truly perpendicular to each other, tighten everything securely.

Bolt the remaining pipe flange to the base plate of the tube saddle—the one described for the simple mounting is satisfactory for this also, as are the rest of the fittings for the tube. Now screw the pipe flange on the threaded end of the declination shaft. In this mounting, the only moving parts are the shafts of the two axes rotating in their bearings, consequently all pipe-fitting joints must be locked in place. You can make sure of this by drilling each connecting part for set screws.

Drill a hole in the side of the tee opposite its side opening. Tap it for a ⅜–20 bolt to which you have attached a 1½-inch valve handle. By turning the bolt down until it bears against the shaft you can lock the telescope in declination in any position you choose.

The counterweight for the long end of the declination axis can be any of the types previously described. If you have used a light tube, a short counterweight shaft is possible. This is an advantage since it gives you more room to move around the telescope.

This completes the fundamental sections of the mounting. You can add refinements later, if you wish, for this type of mounting may be easily and quickly broken down into its component parts and additions readily made to it. These are discussed in a later chapter. This mounting, like the one previously described, may be used on a tripod or on a permanent pedestal.

HOW TO CONVERT A DOBSON TO AN EQUATORIAL

If you add a horizontal platform between the arms of a fork or yoke mounting and then place a Dobson on the platform, the Dobson will track equatorially, once focused on a star. But this would be a cumbersome arrangement, difficult to use.

A much better arrangement was invented by Adrien Poncet of Jura, France,

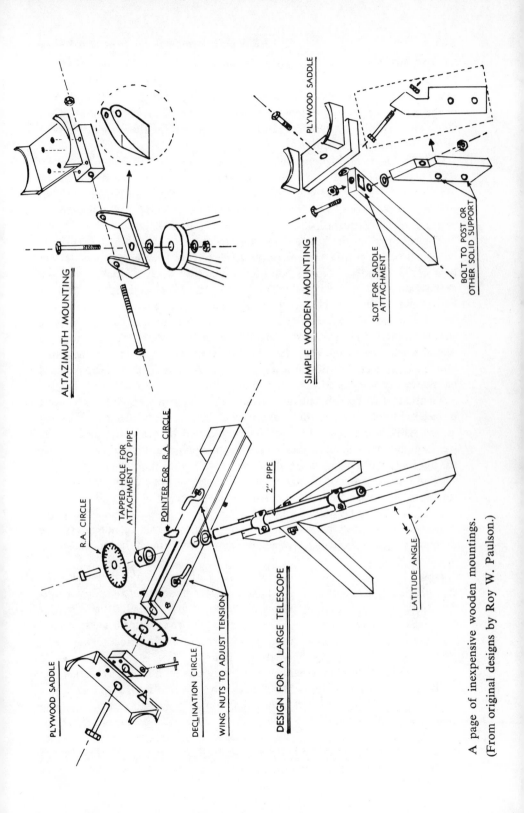

ALTAZIMUTH MOUNTING

SIMPLE WOODEN MOUNTING

PLYWOOD SADDLE

SLOT FOR SADDLE
ATTACHMENT

BOLT TO POST OR
OTHER SOLID SUPPORT

PLYWOOD SADDLE

R.A. CIRCLE

TAPPED HOLE FOR
ATTACHMENT TO PIPE

POINTER FOR R.A. CIRCLE

2" PIPE

LATITUDE ANGLE

DECLINATION CIRCLE

WING NUTS TO ADJUST TENSION

DESIGN FOR A LARGE TELESCOPE

A page of inexpensive wooden mountings.
(From original designs by Roy W. Paulson.)

in 1977 and has been built and used successfully by many amateurs. His platform was mounted on specially angled bearings on a ground board. The idea represents a completely new type of mounting and doesn't fall into any of the other categories of telescope design. The whole arrangement is only a few inches high, and a Dobson can be lifted on or off without disturbing the basic altazimuth structure.

The Poncet is constructed of thick, strong plywood (¾ inch or better) because it must support the entire weight of any telescope placed on its platform. The center of the ground board must lie along a carefully laid out N–S line. The bearings which support the platform are attached to the ground board and the platform itself.

The bearings are the key to the whole structure. On the south end a single bearing acts as a pivot so the platform is free to turn from east to west. It can be a single threaded, headless bolt held in place by wooden blocks, as shown in the diagram. The angle at which the bolt is placed is most important. It must be inclined to the ground board at an angle equal to the latitude of the observer. The inclination of the bolt is referred to in the drawing, and later on in some simple math we shall use, as the Greek letter ϕ. The two blocks that hold the bolt in place must be cut accordingly. A nut is usually embedded in the upper block to hold the bolt firmly in place there, but the lower end of the bolt should be free to move in its socket.

There are two bearings at the north end of the ground board. Each must be free to move so the platform can be tilted. This is accomplished by placing Teflon pads on the ends of two blocks attached to the under side of the platform, allowing them to slide on an inclined plane faced with Formica. As the platform is tilted, one pad slides upward while the other moves down. The slope of the inclined plane is of vital importance to the operation of the Poncet mounting. It must be exactly equal to the complement of the latitude angle, i.e., 90° minus the latitude. Thus if ϕ turns out to be 40°, the inclined plane must be 50°. To test the accuracy of your construction, an imaginary line drawn through the bolt should meet another drawn through the inclined plane at an angle of 90°.

If we have done everything right, the platform should tilt easily and smoothly from east to west and will follow a star if moved at the correct rate. The actual tilting is done by means of a tangent arm. Usually this is simply a thin-bladed hinge attached to the north end of the platform at its center. Pushing on the free end of the hinge moves the platform to the west. The motion must be constant, 15° of arc per hour of time.

A mechanical solution to achieving a steady rate of motion was worked out by Robert E. Cox and Roger W. Sinnott and appeared in *Sky and Telescope* in January, 1977. It consisted of fastening a nut to the end of the hinge and passing a threaded rod through the nut. The attachment of nut to hinge must be temporary so the nut can be spun back to its original position after about an hour's use. When the rod is turned the nut has to move, carrying the end of the hinge along with it. By attaching a synchronous electric motor to the end of the rod and calculating the number of threads per inch needed for a fixed rate, Sinnott built a prototype that provided constant tracking for 53 minutes. Later, Cox worked out the mathematical details for rods of various threads per inch (TPI), the RPM of the motor, and the latitude of the observer φ. They were published in *Telescope Making* in 1980.

Mr. Cox's equations gave two results needed for the construction and use of a Poncet: the distance the nut will travel along the rod in an hour *(A)*, and the distance between the south bearing and the rod *(L)*. To use them you must know the sine of your latitude angle (a table of sines can be found in any high school algebra text). The figure 1436 is the number of minutes in a sidereal day.

A modified form of his equations follows:

$$L = \frac{1436 \times \text{RPM}}{2 \times 3.1416 \times \text{TPI} \times \text{sine } \phi}$$

and

$$A = \frac{1436 \times \text{RPM}}{\text{TPI} \times \text{TPI}}$$

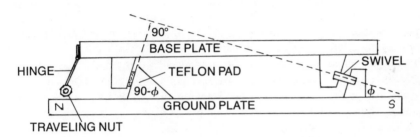

Schematic side view of one form of the Poncet mounting.

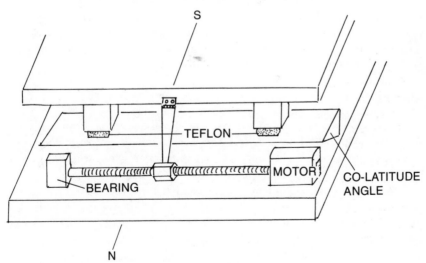

The business (north) end of the Poncet mounting.

Let's use these equations in a hypothetical case: Suppose you have a 1 RPM motor, a threaded rod with 20 threads per inch, and your latitude is 40° north. Then,

$$L = \frac{1436 \times 1}{2 \times 3.1416 \times 20 \times .6428} = 17.8 \text{ inches}$$

$$A = \frac{1436 \times 1}{20 \times 20} = 3.6 \text{ inches}$$

Now you know that the rod must be placed 17.8 inches from the south bearing, and that your platform can be at least this long. You also know that in an hour the nut will travel 3.6 inches along the rod and will turn the platform through 15°. But this is too steep an angle; your telescope is likely to slide off. So you compensate by turning the nut to a position so that the platform is tipped 7.5° to the east. This position will be 1.8 inches from the spot where the platform is level.

Of course, the nut and rod idea is not the only way you can drive a Poncet. You can, for example, use a gear on the bolt at the south end of the mounting, and figure out a way to drive the gear at an equatorial rate. Here is an opportunity to use your ingenuity and imagination. The idea of platform mountings is relatively new; much can be done with it.

The Poncet mounting is an excellent supplement for any optical instrument not having its own tracking system. But it has its limitations. You will have noticed that it has no provision for tracking in declination, important in observing the moon and planets. Although it works very well between latitudes from 35° to 65°, at low latitudes the inclined plane becomes too steep to be efficient; for high latitudes it is too shallow. Exposure time for photography is limited to an hour. Finally, it does not work very well with objects near either horizon. But it is really excellent for those near the meridian and along the celestial equator, where there are many striking objects to be viewed.

Many of the difficulties with the Poncet are minor, in spite of what has been said above. And it is easy to make and fun to use.

FINDERS

We digress for a moment at this point to talk about an important addition to the telescope, the finder. Ordinarily we should discuss this along with other additions to the telescope, but since a finder affects the balance of the mounting we include it in this section.

The field of view of your telescope, using a low-powered eyepiece, will be about the size of the full moon, or ½° of arc. The moon is always easy to find with a telescope; its diffused light spreads over several degrees and can be seen in the edge of the field long before the moon itself comes into view. But what if you are looking for a faint object such as the Ring Nebula in Lyra, surrounded by no helpful glow?

You have several means of finding such an object. You can point the telescope toward the constellation in which it occurs, find the brightest star near the object you are looking for, and then search in the area around it. But this is not very satisfactory and you may look for a long time before picking up the elusive object. A second choice is to equip the telescope tube with some sort of gunsight arrangement lined up with the optical axis of the telescope. Paint the sights with luminous paint or attach patches of the luminous material which is used as warning devices for automobiles and bicycles. Now you can point the telescope much more accurately, although you are still limited to naked-eye (6th magnitude) guides in the vicinity of the object you are looking for. Still another choice is equipping the polar and declination axes with setting circles. Now all you need to do is to pick the coordinates of the object from a star chart, adjust the readings for right ascension to your local sidereal time, and move the setting circles to correspond to them (see chapter 14 for a description of setting circles). Then, when you look in the eyepiece, the object of your search looks right back at you. This is by far the best method, and you will eventually want to equip your telescope with setting circles. But even if you are fortunate enough to have a good pair of circles on your telescope, there will be many occasions when you will find it easier and quicker to use a finder telescope.

The finder is simply a low-powered telescope of fairly wide field whose optical axis is in line with that of the main telescope. Because of the wide field, a faint object is easily located and centered in the finder eyepiece. Once this is accomplished, the object also appears in the eyepiece of the main telescope.

The field of the finder should not be more than $3°$ or $4°$. Larger fields make locating any given object easier but they also increase the difficulty of getting the object into the field of the main eyepiece. A slight deviation from the center of the field of the finder will remove the object completely from the field of the telescope, even if the axis of both finder and telescope are exactly in line. Many telescope makers fail to realize this and make finders of $10°$ or $12°$ field. A simple illustration shows why this is not good practice. Suppose the field of the finder is $10°$ and that of the telescope $\frac{1}{2}°$. If the deviation of the object to be found is only $\frac{1}{19}$ of the total field away from the center of the finder, it will not appear in the main eyepiece at all. Consequently, the field of the finder should not exceed 6 times that of the telescope when the lowest power telescope eyepiece is used.

It is also a good idea to equip the finder with cross hairs or a reticle of some kind, since these are a great help in centering an object in the field. If you use a positive eyepiece, place the cross hairs just in front of the field lens, that is, in the focal plane of the eyepiece. A reticle is a thin piece of glass upon which lines are printed or engraved and which serves the same purpose as cross hairs. Like the latter, a reticle must be placed in the focal plane of the eyepiece. If your finder eyepiece is a negative lens combination, you must place the reticle or cross hairs at some point between the eye lens and the field lens because this is where the focal plane of this type of eyepiece is located.

Possibly the least troublesome and expensive way to obtain a finder is to buy an inexpensive refracting telescope from war surplus stock. There are many of these still available and some of them have the proper characteristics—6 power and a field of 3° or 4°—which are needed for a good finder. They have the disadvantage, in many cases, of too much weight and therefore require heavy counterbalances. Of course *any* finder must be counterbalanced; but, since you have made a strong, light tube for your telescope, you will not wish to add needless weight to it. A further drawback is that the war surplus finders usually produce erect images, while your telescope produces an inverted image. The stars will appear in the finder in the same positions they occupy on the star chart, while those in the main eyepiece will be reversed. Most people find it easier if the reversal occurs between chart and finder—where the field is larger and orientation more simple—than between finder and main telescope.

MAKING A SIMPLE FINDER

You can make an excellent finder from war surplus lenses. A 1½-inch achromat with a focal length of 6 inches can be used with a 1-inch focal length eyepiece to produce a power of 6 and a field of 6°. This is a little too wide, but you can decrease it by introducing a stop in front of the field lens to cut the field down to whatever you wish. The achromat costs in the neighborhood of $7–$10 and you can make an eyepiece for as little as $15 (see chapter 11). Eyepieces for finders are not critical equipment; you can use one which would not be at all satisfactory in the telescope itself.

Make the mounting for the finder from two rings cut from a piece of heavy-walled pipe. Mount these rings on wood blocks at least 2 inches high to provide adequate clearance for the eye. Drill and tap each ring at three points 120° apart for set screws to support the finder. The set screws will supply all the adjustment you need to align the axis of the finder with that of the telescope.

Place the finder mounting near the eyepiece of the telescope, close enough so that it is easy for the eye to be shifted quickly from one eyepiece to the other, but not so close that it interferes in any way with easy access to the main eyepiece. Most observers prefer to place the finder on the tube at a distance of 45° from the main eyepiece holder. With the finder in position, balance the tube as described at the end of this chapter. After you have lined up, or collimated, the optical train of the telescope, set it upon a bright star and center the same star in the finder by means of the set screws. Then test the alignment by trying it on other stars in different parts of the heavens.

BALANCING THE TUBE

In order to work smoothly and easily, the telescope must be perfectly balanced. This is one respect in which many amateur telescopes are inferior to those made professionally (and even here there is often much to be desired). If the moving parts are tightened, most telescopes may be moved to any position with reasonable confidence that they will remain there. Too many of us are satisfied with this condition. Achieving "balance" by this method has two serious drawbacks. Tight fittings are subject to excessive wear on moving parts with the result that they eventually loosen and have to be tightened again. The end effect is eventual replacement or repair of these parts. Even more serious is the fact that poor balance is always a source of vibration in the instrument as a whole because the tight fittings transmit vibrations from one part of the telescope to adjoining sections. For this reason each part of the telescope should be balanced carefully with respect to its own individual motion, and then balanced again with respect to its part in the motion of the whole instrument.

Radial Balance

All the parts of the tube should be assembled in their final position and the entire unit balanced in two directions, radially and longitudinally. Achieving tube equilibrium radially may, at first glance, seem to be an unnecessary refinement, but it is an absolute essential if the tube is to rotate within its saddle so the eyepiece may be easily reached. If the eyepiece and finder are oriented away from the mounting for one position of the telescope and toward it for another, it is obvious that the counterweight at the other end of the declination axis will not take care of both positions as far as balance is concerned. You could, of course, change the position of the counterweight for varying positions of the tube, or you could add a supplementary counterweight to the main one, thus providing for all possible positions of the tube. But this would involve making two adjustments for a new telescope position, one to adjust the eyepiece position and the second to find the new point of equilibrium. Much simpler than this is adding weights at a strategic point on the tube itself so that it is balanced in *any* position.

Since the primary mirror and diagonal holder are centered in the tube, the only other parts which can produce imbalance are the eyepiece holder and finder. These, as we have already said, are usually about 45° apart. The result is a concentration of weight on one side of the tube which may be compensated for by placing an equal concentration on the opposite side. But exactly where the weight should be placed and how much should be put there is a difficult problem for most amateurs to solve. If the weights of the finder and eyepiece holder are known exactly, a trigonometric solution of the problem may be obtained.[1] Most amateurs, however, would prefer a less complicated approach. Here is one which is practical for our purpose.

Remove the cell from the tube, since it has little effect upon the radial balance. Cut two pieces of 1×2 a little longer than the inside diameter of the tube. Round off their ends so that each will make a snug fit across the ends of the tube when they are lightly forced inside. These will serve as a support for the suspended tube. Find the center of each piece by drawing intersecting diameters, remove them and drive a stout nail at each point. Replace them in the same positions and support the tube on "V"-shaped chucks at each end, using the nail heads as pivots. The tube is now free to rotate around its longitudinal axis.

Find a point about midway between the eyepiece and finder if they are the same weight or proportionately closer to the heavier if they are of different weights. For this purpose a rough estimate of the weights is good enough. Assemble a few objects (short pieces of strap iron, a handful of 20-penny nails, etc.) whose combined weight is about equal to that of the finder and eyepiece together. Use masking tape to attach them to the side of the tube at a point *opposite* to the one found between the eyepiece and finder. Build up an accumulation of these objects until the tube remains in any radial position in which it is placed. If you can't make the tube balance in all positions, change the location of the weights slightly and try again. You will eventually find a combination of weights which will achieve a balance. Remove the weights and weigh them. Cast an equal weight of lead in the bottom of a 2-inch tin can. Mount this counterweight on a rod placed parallel to the tube at the position decided upon. Using a rod makes it possible to change the arrangement later if you decide to add anything else to your tube.

Longitudinal Balance

This is a relatively simple operation compared to achieving radial balance for the tube. Your task here is to find the center of gravity of the tube and to attach it to the saddle in such a way that the center of gravity is on a direct line with the longitudinal axis of the declination shaft. The simplest procedure for this is to make a sling and balance the tube and all its contents—including the cell—at various places until you find an equilibrium point. Of course the center of gravity is *inside* the tube, while you are balancing it from the *outside,* but for all practical purposes this method will work well. Mark the point of balance on the tube lightly with a pencil; then turn the tube over and balance it again. The two points should be exactly opposite one another, if they are not you must find a series of points around the tube and take their average.

[1] R. E. Cox, *Sky and Telescope,* November and December 1958.

Balance the saddle and then the tube in the declination axis as described earlier (page 225). This method of balancing a telescope is admittedly an approximate one, but it will bring the instrument to balance within very close limits. Any remaining imbalance can be corrected by adjusting the friction clutch on the polar axis or the locking device on the declination shaft. You may be pleasantly surprised at the small adjustments you will find it necessary to make.

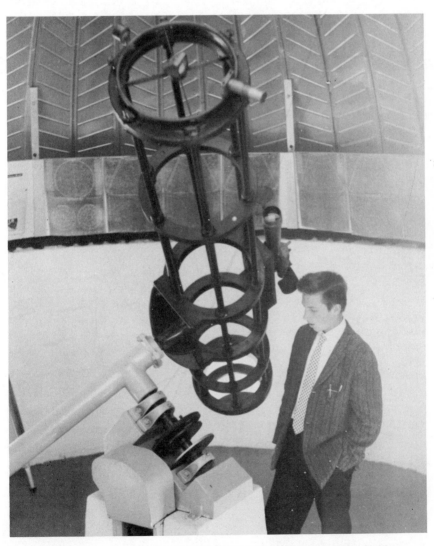

Open-tube design for a long-focus (f/10) 12-inch telescope. The drive motor is concealed by the housing at left. Shafts are 2¼ inch, contained in 4-inch pipe. The dome was made from a 17-foot silo top.

Adjustments and Additions

ADJUSTING THE TELESCOPE

THE HEART AND SOUL of the telescope are the three basic optical elements: mirror, diagonal, and eyepiece. All three may be optically perfect; but, unless they are lined up properly, or collimated, with respect to each other, the telescope will not produce a sharp image no matter how carefully focusing is done. This is the fundamental operation in adjusting your telescope. But for the highest performance in all respects you must go beyond the collimation of the optical parts. You must be sure, as well, that the axis of the mirror is at right angles to the declination axis, and that the polar axis is truly parallel to the axis of the earth. Consequently, adjusting the telescope consists of three steps:

1. Collimation of the optical train
2. Adjustment of the declination axis
3. Adjustment of the polar axis

COLLIMATING THE OPTICAL TRAIN

Much has been written about this aspect of aligning the telescope and the subject has been treated as though it were extremely complex. As a matter of fact, there is no need for the cross hairs, pinholes, diaphragms, screens, or any of the other paraphernalia which has been thought necessary in the past. We need only one piece of supplementary equipment: the eye cap from a high-powered eyepiece so that we can be sure the eye is centered on the draw tube of the eyepiece holder. Nor do we need to remove the telescope from its saddle and take it indoors for the testing procedure. The best objects for testing alignment are found outdoors; a couple of stakes in the ground and a 3rd-magnitude star will do very nicely.

The first step is rough collimation. Remove the eyepiece from the telescope and replace it with the cap from a high-powered eyepiece. If this is not available, any covering with a small hole which can be centered over the draw tube of the eyepiece holder will do. Check first to be sure the eyepiece holder is perpendicular to the *side* of the tube. If your tube is perfectly straight, this is also an adequate check that the eyepiece holder is also perpendicular to the *axis* of the tube. The side of a square laid along the tube must be parallel to the edge of a ruler held against the inside edge of the draw tube.

Look through the hole in the direction of the diagonal mirror. If this has been properly placed, you will see in it an image of the end of the tube containing the main mirror. But if the diagonal is tipped out of line, you will see one side of the tube and only part of the main mirror. In this case, adjust the positioning screws on the diagonal until the edge of the main mirror appears equidistant at all points from the edge of the diagonal. Incidentally, this is also a check on the proper size of the diagonal. If you have figured correctly, the image of the primary mirror should almost cover the diagonal; that is, the edge of the primary should appear to fall just inside the edge of the secondary.

Now look for the image of the diagonal itself in the main mirror. It should appear in the center; if it does not, you can adjust the position of the cell by turning the supporting screws at the back of the mirror cell. If the reflection of the diagonal is displaced to either side, advancing the screws on that side of the primary mirror will shift it toward the center. You may, if you wish, check on the centering of the diagonal image by stretching threads across the face of the primary and bringing the image to their intersection. Some workers consider this an unnecessary refinement since the eye alone is capable of detecting small deviations from a central position.

When all optical elements are in line you will see the following as you look through the hole in the draw tube.

1. The diagonal mirror itself centered in your field of view.
2. An image of the primary nearly filling the surface of the diagonal mirror.
3. A reflected image of the diagonal centered in the primary mirror.
4. Your own eye, seen by double reflection, centered in the diagonal mirror.

For most purposes, this preliminary lining up of the optical train will be entirely satisfactory. A slight displacement of the elements is not usually apparent at low powers when the eyepiece is focused. But at higher powers, slight defects in placement will cause coma (the flaring of a star image to one side) or even double images. So you proceed to more exact collimation by using a 3rd-magnitude star as a measuring device.

Use a low-powered eyepiece, choose a star somewhere overhead to be sure that the mirror will remain seated on its supports, then move the eyepiece out of focus away from the mirror. Coma, diffraction effects, and any astigmatism present will make their appearance at this point by a flaring of the image and a series of partially broken diffraction rings on the side of the image on which the flaring occurs.

Making adjustments with the telescope pointed overhead is a very difficult operation for one man since he must be in two places at once—looking through the eyepiece and adjusting the set screws at the back of the cell. So it is best to employ an assistant to adjust the screws for you while you give directions from your position at the eyepiece. Most of the adjustment will be on a trial-and-error basis and is done by turning any of the three screws which seems to correct the displacement. In general, the image must be displaced *toward* the flare on the star. When you have succeeded in producing the roundest image possible, change the eyepiece for one of higher power. The diffraction rings which surround the star image are more evident under higher powers, inside and outside of focus. The eyepiece should be moved out of focus and final adjustments —these will be very small ones—are made until you have succeeded in centering the star in its diffraction rings.

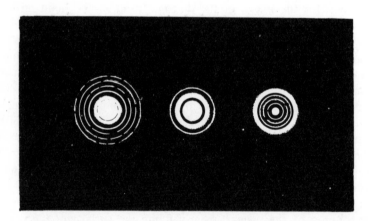

Star diffraction patterns. *Left,* eyepiece inside focus; *center,* at focus; *right,* eyepiece outside focus.

Testing the optical system for collimation by using a star should preferably be done on a night when there is little turbulence in the atmosphere. Such a night is hard to find, and some of the defects noted may be due to atmospheric disturbances rather than in the mirror. A good rule is that any defect you observe which appears to shift position is probably due to causes outside the optical system. But if the defect does not move, or disappears and reappears in the same position, it is caused either by irregularities in the mirror surface or in the way the optical system is lined up.

LINING UP THE DECLINATION AXIS

The alignment of the optical system in the tube is fundamental to good visual observation of heavenly objects, but if photographic work is also an objective, and if setting circles are to be added to the system, both axes of the telescope must be adjusted to their optimum positions.

We start with the declination axis. If space permits, drive two stakes in the ground 200 feet apart. Place a nickel-headed thumbtack at the top of each facing surface of the two stakes. Halfway between, drive a third stake and place a thumbtack on its top exactly in line with the other two. Move the telescope to a position where the turning point of the tube around the declination axis is exactly over the thumbtack on the center stake, using a plumb bob for accuracy. To do this, turn the telescope on the polar axis until the declination axis is horizontal. Check this position with a carpenter's level.

Point the tube toward one of the stakes until you pick up the thumbtack in the eyepiece. Swing the tube in declination through 180° until the tack on the other stake comes into view. If you are very fortunate, each tack will be centered in turn in the eyepiece as you swing the tube. But the chances of this happening on the first trial are not very good, so you will probably have to make some adjustments.

Start by moving the whole telescope to right or left to bring *one* of the tacks into the *center* of the field. Swing the tube and pick up the tack on the opposite stake. This will not be centered in the field, of course, and you must estimate how far from center it is located. Loosen the restraining band around the tube and place shims at one end of the saddle until the tack has apparently moved toward the center, half of the distance noted above. Reverse the tube again and once more shift the whole telescope until the first tack is centered. When you return to the second tack it also should be centered, or very close to it. Repeat the whole process until both tacks remain centered when the tube is reversed. When you have accomplished this, you have succeeded in orienting the mirror at right angles to the declination axis.

ADJUSTING THE POLAR AXIS

If the telescope rests on a pier the polar axis can be permanently adjusted by making sure that the base plate of the mounting is perfectly level in its position on the pier and then orienting the axis on the North Star. If the telescope is mounted on a tripod, the plate of the tripod must be leveled each time the telescope is used. Assuming that this will be done, you can make a permanent adjustment by lining up the axis with the base plate, using the North Star as above.

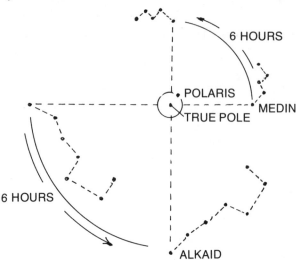

Using the circumpolar constellations to line up the polar axis. When the line joining Alkaid (in the Big Dipper) and Medin (in Cassiopeia) is vertical, Polaris indicates true north. When the line is horizontal, the altitude of Polaris equals the latitude of the observer.

But adjusting the axis to the apparent position of the North Star is like lining up the sights of a gun on a shifting target, for Polaris does not always represent the north point of the heavens. True north is an imaginary point around which Polaris revolves at a distance of almost one degree. Hence the North Star indicates *true north* only when it is directly above or directly below the imaginary pole. And it indicates the true *elevation* of the pole only when it is directly east or west of this imaginary point. Nevertheless these conditions are sufficient for us to line up the polar axis very accurately. Fortunately, we are provided with a set of indicator stars which tell us exactly when Polaris is in a true northerly position and also when it marks the true elevation of the celestial pole. Two of the circumpolar constellations provide the indicator stars, Schedar in Cassiopeia (see diagram) and Mizar in the handle of the Big Dipper. A line drawn between these two passes almost exactly through Polaris. When the line appears vertically in the sky, with Schedar on top, Polaris is above the true pole and in line with it. Six hours later, when the line is horizontal with Schedar to the left (west) as you face north, Polaris is at an angle above the plane of your horizon which is exactly equal to your latitude.

The adjustments to the polar axis are therefore made in two steps six hours apart. In the first step, adjust the polar axis horizontally by clamping the declination axis in a horizontal position and then turning the mounting on its base plate to right or left until Polaris is centered in the field. When Polaris has moved to a horizontal position, six hours later, clamp the declination axis vertically and change the angle of the mounting on its base plate until the star is again centered. This may be done by loosening the main supporting bolt. Each of these steps should require very small adjustments to the position of the mounting if you were careful, in the original construction of the mounting, to measure the polar axis angle to fit your latitude angle.

These three steps—collimating the optical axis and adjusting the declination and polar axes—have tuned the motions of the telescope to the corresponding motions of the stars. Now to find the stars, especially the faint ones where an optical finder is of little use, we shall need some setting circles. Finally, if we wish to follow them for photographic purposes we shall require a driving mechanism of some sort.

SETTING CIRCLES

In talking about the purpose of the declination and polar axes, we mentioned the celestial sphere and the geography of the heavens. Now we must go into this subject a little further in order to understand the application of setting circles to the position of the stars on the celestial sphere. For purposes of astronomy, we can classify all celestial objects into two main classes, those whose positions are fixed in relation to each other, and those which move among the fixed objects. In the first class are stars, nebulae, star clusters, and galaxies. In the second group are the planets, the moon, and the sun. The members of this group—the solar system—move not only in relation to the fixed stars but also in relation to each other. Both classes have one thing in common; their motion is fixed and regular as seen from the earth and their positions for any given instant of time can be calculated in advance, as we shall see.

The north-south position of a star, or its declination, never changes— or changes so little that for practical purposes we can consider it fixed. To find the declination of a star we consult an ephemeris or find it on a good star chart (such as the Skalnate Pleso charts) and locate the star relative to the scale on the side of the chart, just as we locate the latitude of an object on earth by a similar scale on the side of the map.

Ninety degrees north declination (the north is sometimes designated by a + sign) is the north pole of the celestial sphere; 90° south declination (similarly designated by a − sign) is its south pole. The celestial equator, or equinoctial, is at 0° declination. The position of a star in declination, as seen from various points on the earth, can best be illustrated by examples. Suppose a star is located somewhere on the equinoctial, or at 0° declination. Seen from the equator, it would appear to pass directly overhead in its motion across the sky. Seen from a position in the northern hemisphere, it would pass to the south of the observer. In short, there is a definite relationship between the location of the observer on the earth and the position of the star as it passes across the heavens, and it is possible to establish the relationship by a scale attached to the telescope.

Suppose we construct a semicircle and at the midpoint of its edge place a point that we call 0°. Now we place the semicircle in a north-south position and tip its base until the 0° mark points to the celestial equator. If we graduate the half of the semicircle which points north into intervals from 0° to 90°, these will correspond to points in north declination. Points on the other half of the semicircle will measure south declination. But how can we find the point in the sky which represents the equinoctial in the first place? This needn't cause much trouble, because it must be as many degrees above the southern horizon (if we live in the northern hemisphere) as the complement of our latitude, which is simply the latitude subtracted from 90°. Put more simply, if we lay the base of the semicircle along the polar axis of the telescope, the 0° point will lie on a line parallel to one drawn between the center of the earth and a point on the equinoctial. To point the telescope in declination, we need only line up the tube with this scale.

It is simpler, in practice, to attach the scale to the declination axis so that it rotates with the axis and take readings by means of a pointer attached to a fixed part of the telescope. Furthermore, there are two positions that the declination axis may assume in the German mounting and still permit the telescope to point to the same part of the sky. If the weight end of the declination axis is directed east, the telescope itself will be to the west of the pier or tripod, and vice versa. So it is easier to produce two scales, even though one is all that is needed. A single scale *could* be read in both positions of the declination axis, but in one position the scale will be erect and in the other, inverted. Providing two scales simply means using the whole circle instead of half of it and using either half as the position of the declination axis dictates. In case this is hard to visualize, cut out a rough circle of cardboard, attach it by means of a hole through its center to the declination axis, and then turn the telescope to various positions around the pier, noting that the card reverses for opposite sides of the pier.

Making and Mounting the Declination Circle

In making any setting circles, whether for declination or right ascension, don't make the mistake of being niggardly in size because the graduations, being close together, will be hard to read. A diameter of 6 inches is the smallest tolerable size; since you are making them yourself you can choose any size which will fit your telescope.

Dividing a circle into a number of equal units is a problem because it is difficult to transfer linear measurements to a curved surface. If you can find a large, accurate protractor you can achieve a fair degree of precision in laying out the scale by placing the protractor in the center of your circle and extending the lines to the edge. A better method is to make the circumference of the setting circle of a length in which inch measure-

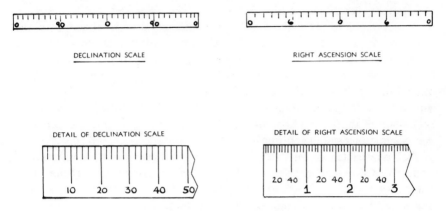

Scales for setting circles.

ments can be converted into degrees. A good circumference for this purpose is 22½ inches; this makes each inch equal to 16°, and the ⅛-inch markings on an ordinary ruler now equal 2°. The field of the telescope with your low-power eyepiece is ½°; by estimation between the 2° marks you can easily come within this range. A circumference of 22½ inches requires a diameter of 7⅙ inches.

Cut a disk of ¾-inch plywood 7¼ inches in diameter. Turn it on a lathe or hand sand it until a uniform diameter of 7⅛ inches has been achieved. Your scale will make up the difference in diameter since it will consist of a piece of plastic superimposed on the circumference of the disk. Cut a hole the exact size of the declination axis in the center of the disk.

Cut out a strip of polyvinyl plastic, 10 gauge or heavier, 30 inches long and the width of the disk edge. Lay a rule along the strip and mark off the ⅛-inch graduations of the rule on the plastic. Use a scriber to scratch these graduations into the strip, making every fifth scratch a little longer than the others. Fill in the scratches with white paint. Mark the long intervals in units of 10°, from 0° up to 90° and back down again, using the largest type you can get on the scale. Rubber stamp numbers, using white ink or paint, are excellent for this purpose. When you have finished, you should have a strip 22½ inches long, with two units of 0° to 90° to 0° marked on it. Place this around the disk. If the ends overlap, build up the disk by gluing strips of thin paper around it until the ends of the scale just meet. If the disk is too large, sand it until the strip fits. Now paint the circumference of the disk with glossy black enamel. Glue the strip around the disk, using any transparent glue available. Make sure the strip is secure by using half a dozen small flat-head screws placed at equal intervals around its circumference. The white numbers against the black paint make the scale easy to read. Spray the whole unit with transparent plastic for permanence.

Attach the finished disk to a collar which in turn is attached to the declination axis with set screws at the point where the shaft emerges from the bearing at the end opposite the telescope tube. Make an indicator by drilling a ½-inch circle in a piece of heavy aluminum sheet, as shown in the diagram on page 215. Solder a wire across the circle, then attach a thin piece of luminous adhesive to the wire. Attach the indicator to the tee of the declination shaft so that the luminous wire lines up with the graduations on the setting circle.

Pick out a star at an overhead position, look up its declination, and center the star in the eyepiece of the telescope. Turn the setting circle on the shaft until it is set to the known declination, then tighten the set screw. Reverse the direction of the declination axis by moving the telescope around the pier, center the star once more in the eyepiece, and check the reading on the other half of the circle to be sure it matches the first reading. Test the whole arrangement further by moving the telescope to the declination of another known star. If you move the telescope in right ascension (around the polar axis) this second star should appear in the field. If the readings differ on opposite sides of the circle it is because the circle is not centered on the shaft or, if this is not the case, because you have gone astray in placing the scale on the disk. You can cure the first possibility by removing the disk and changing its center hole, making up the enlarged hole by shimming around the inside or by using more set screws. If the scale is wrong, however, you must either make a table of corrections for one side or the other or do the whole thing over again.

The Right Ascension Circle

Lines drawn from the poles of the earth which pass through the equator are called meridians and measure longitude. Corresponding lines on the celestial sphere are known as celestial meridians, or hour circles. The one which passes directly overhead is known as your local meridian. The earth system makes use of the meridian which passes through Greenwich, England, as a starting point, and longitudinal measurements are made east or west of this line. Astronomers choose as a base reference line in the heavens the meridian which passes near the constellation Aries, the Ram. The exact point through which this line passes, called the vernal equinox, or the First Point in Aries, is the intersection between the celestial equator and the ecliptic. The ecliptic is the path followed by the sun and planets as they appear to travel around the earth.

For purposes of star measurement it is convenient to think of the earth as a stationary body with the celestial sphere revolving around it. But here we must include a new concept of time for, unlike the sun, the celestial sphere appears to revolve around the earth in 23 hours 56 minutes instead of an even 24 hours. This means that any star will cross over the same spot on earth 4 minutes earlier each night if we measure its passage in solar time. The time system based on the stars is called sidereal time, and the sidereal day starts when the First Point in Aries (which we shall refer to simply as Aries from now on) is directly over the observer, or crosses his local meridian. This is called zero hours, sidereal time, and the sidereal day is divided into 24 sidereal hours which are numbered from 0 to 24, instead of two groups of 12, as in solar time.

The positions of the stars are measured east of Aries. As they cross the local meridian in their apparent westward journey, their position is measured in the *time* which has passed since Aries crossed the local meridian. Thus a star which appears one hour later than Aries (or is one hour east of it) is said to have an hour angle of one hour, and a

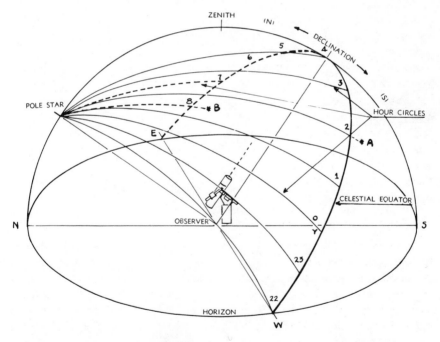

In this diagram, the meridian of the observer has a right ascension of 4 hours and the sidereal time is therefore also 4 hours. Since star *A* has a right ascension of 2 hours, it will be found at an hour angle 2 hours *west* of the observer. Similarly, since star *B* has a right ascension of 8 hours, it will have an hour angle 4 hours *east* of the observer.

star which crosses an hour earlier than Aries has an hour angle of 23 hours. The same relationship is indicated on star charts, and the eastward hour angle from Aries for any star is called its right ascension, or RA. If the coordinates of a given star are, say, RA 16h 30m and declination N 60°, the star will cross any local meridian 16 hours 30 minutes after Aries has passed and its elevation above the celestial equator will be 60°.

Finding Local Sidereal Time

Now if we can find some way of connecting sidereal time with solar time we can use the relationship to find out where to look in the sky for any heavenly body. To do this we must have a starting point on earth from which to measure both systems. The point and time chosen are Greenwich, England, at the beginning of the solar day, or 0 hours. The table in the appendix (Table II) headed "Sidereal Time at 0 Hours for the Meridian of Greenwich" provides the base point for our calculations. This table shows how many hours of sidereal time have passed since Aries crossed the Greenwich meridian. For example, at 0 hours on April 23, Aries had passed by 14 hours and 1 minute earlier. So we say that the sidereal time is 14 hours 1 minute. But suppose we wish to find the sidereal time for Greenwich at 6 o'clock in the evening instead of at 0 hours?

First, we must convert the 18 hours of sun time into a corresponding period of sidereal time. We said earlier that 24 hours of sidereal time equals 23 hours 56 minutes of solar time which, when reduced, indicates that a sidereal hour must be 10 seconds less than a solar hour. Therefore to make the conversion we must add 10 seconds to each hour. For convenience's sake, we have included a table (Table I) in the appendix by which any solar period of time may be converted into the corresponding sidereal time. Using Tables I and II for the above problem we have:

Sidereal time at 0 hours (Table II)	14 h	1 m
Hours since 0 hours	18 h	0 m
Correction (Table I)		3 m

Sidereal time at 6 P.M., April *23* = 32 h 4 m, and, throwing out the extra 24 hours, the sidereal time must be 8 h 4 m, April *24*.

But this is only for an observer at Greenwich. Suppose you are somewhere else on the globe, say Philadelphia. In such a case we go through the following steps:

1. Convert the local time at Philadelphia into the local time at Greenwich according to the proper time zone. To keep in step with the sun, time on our globe is divided into 24 zones, each 15° wide, starting from Greenwich, England. The time zones are numbered

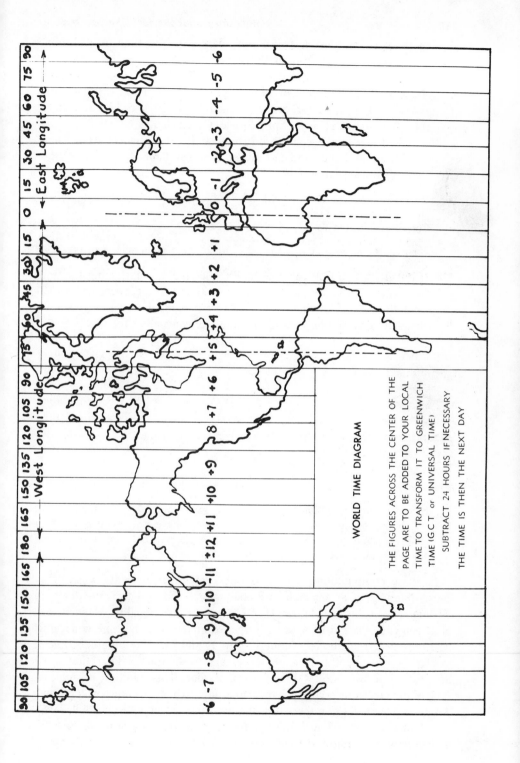

WORLD TIME DIAGRAM

THE FIGURES ACROSS THE CENTER OF THE
PAGE ARE TO BE ADDED TO YOUR LOCAL
TIME TO TRANSFORM IT TO GREENWICH
TIME (G C T or UNIVERSAL TIME)

SUBTRACT 24 HOURS IF NECESSARY
THE TIME IS THEN THE NEXT DAY

both eastward and westward (see diagram). The time in each zone is an hour earlier than in the zone immediately to the east of it. Thus Philadelphia, whose longitude is 75° 8′ 57″, falls in zone 5. Now if the local time in Philadelphia is 9:30 P.M., April 23 (or 2130 using a 24-hour clock), it will be 2130 + 5 hours = 2630, or 0230 the next day, April 24, at Greenwich.

2. Convert this Greenwich Time (usually called Greenwich Mean Time, or Universal Time) into Greenwich Sidereal Time. Using the system described above, we have:

Sidereal Time for 0 h, April 24 (Table II) 14 h	5 m	
Hours since 0 hours	2 h	30 m
Correction (Table I)		25 s

Sidereal Time at Greenwich	16 h 35 m	25 s

3. Convert this to sidereal time at Philadelphia by subtracting the time interval for the exact difference in longitude between Greenwich and Philadelphia, which is 75° 8′ 57″. Table III in the appendix, headed "Conversion of Arc to Time," is convenient for this.

	h	m	s
for 75°	5		
for 8′		32	
for 57″			3.8
	5	0	36

Subtracting this from the Greenwich Sidereal Time:

16 h	35 m	25 s
5	0	36

11 h   34 m   49 s, or 11 h 35 m, which is the Sidereal Time at Philadelphia for 9:30 P.M., April 23. A little thought will make it evident that the sidereal time we have just found is identical with the hour angle (see page 237) of Aries, the only difference being that it is expressed in units of *time* rather than *angle*.

In these computations there are only one or two points which may cause difficulty. If at any point we encounter a time which is greater than 24 hours, we subtract 24 from it. Or if we are in a locality which is *east* of Greenwich, instead of *west* as in the example used, the time zone must be *subtracted* from the local time in step 1, and the time difference *added* in step 3. Finally, if it becomes necessary to subtract a later time from an earlier one, we add 24 hours to the earlier. This addition or subtraction of 24 hours, of course, changes the date, which we must watch when we enter the tables for the sidereal time for 0 h Greenwich Civil Time.

All this can be put into tabular form, such as is suggested below, together with an example for its use.

FORM FOR FINDING LOCAL SIDEREAL TIME
(Hour Angle of Aries)

		Example[1]
Date	....	Jan. 5, 1959
Your longitude	....	* 74° 37′ W
Zone time	....	2300 (or 11 P.M.)
Zone difference	....	* 0500
(+ if west:		
— if east)	....	2800
24-hour correction	....	− 2400
(if necessary)		

Greenwich Mean Time	....	0400, Jan. 6.		
Hours since 0 h	....	4 h	0 m	0 s
Correction (Table I)	....			40 s
Total	....	4 h	0 m	40 s
Sidereal time for 0 h				
(Table II)	....	7 h	0 m	0 s
Greenwich Sidereal Time	....	11 h	0 m	40 s
Longitude difference				
(Table III)	....	* 4 h	58 m	28 s
(− if west: + if east)				
Local Sidereal Time	....	6 h	2 m	12 s

After the sidereal time for your locality has been found, the final (and easiest) step is to use it for the location of any heavenly body. Using the example already given in the tabular form above, suppose we wish to find the great nebula in Orion at 11 P.M., local time, January 5, 1959. A star chart gives the information RA 5 h 33 m and declination − 5° 25′ for this object. We have already found the sidereal time to be 6 h 2 m, which means that Aries is 6 h 2 m *west* of our meridian. But if the nebula is 5 h 33 m *east* of Aries, it must be 29 m *west* of our meridian. This is called the local hour angle. Therefore we must point the telescope this much west of our meridian, elevate it in declination to − 5° 25′, and the nebula will be in the field of the telescope. The computations are:

Our meridian	6 h	2 m	East of Aries
Orion Nebula	5 h	33 m	East of Aries
Therefore the Orion			
Nebula must be		29 m	West of our meridian

Another example: a spiral galaxy in the constellation Leo has a RA of 10 h 44 m, Dec. + 12° 05′. Where shall we point the telescope?

[1] All figures marked with an asterisk will be constant for your location; they may be printed on the form for general use unless you move to a new location.

Galaxy in Leo	10 h 44 m	East of Aries
Our meridian	6 h 22 m	East of Aries
Therefore the galaxy		
in Leo must be	4 h 22 m	East of our meridian

These two examples demonstrate that, if the RA of the object is *less* than the local sidereal time, the object will be to the west of the local meridian; but, if the RA is *greater,* the object will be to the east.

Finally since 12 hours represents 180° of the heavens, it is evident that an object whose hour angle is more than 6 hours different from the sidereal time will not be visible at all.

At long last we have reached the point of this protracted discussion, which is that if we can build a scale on the telescope that will measure hour angle, we can use it to locate the east-west position of a celestial object at any point in time. The scale will be located on the polar axis and will measure hour angle east or west of the local meridian. We call it the RA setting circle.

Constructing the RA Setting Circle

Proceed exactly as with the declination axis setting circle, using the same dimensions (7⅛ inches). This circle will be marked off in hours and minutes instead of in degrees. Each ⅛-inch mark on the ruler represents a time unit of 8 minutes instead of 2°, as on the declination axis. Since 8-minute intervals are inconvenient for measuring hours, it is better to use 4-minute intervals, each represented by 1/16 inch. Draw a diameter on the circle; mark the opposite ends of the diameter 0 hours. On another diameter at right angles to the first, mark the opposite ends 6 hours. Now cut out a strip of plastic of the same length and width as that of the declination circle. Using an accurate rule, divide the strip into marks 1/16-inch apart, each of which will represent 4 minutes of time. One hour equals 15 of these intervals, so you will need to make a long line to represent hours at every fifteenth line. An intermediate-length line placed every fifth interval now represents 20 minutes. Number the 20-minute and hour intervals, using different-sized numerals. The whole scale covers 24 hours in 4 units of 6 hours each, as shown on the diagram.

Attach the plastic strip to the 7⅛-inch disk so that the 0- and 6-hour marks fall at the ends of the diameters. Now mount the setting circle on the polar axis shaft at a convenient place. The lower end of the shaft, where it protrudes from the bearing is usually the best spot because you will not have to take the mounting apart to slip the circle on the shaft. The set screw which holds the setting circle in place should be a thumb screw so that you can change the setting easily.

The setting circles described here are somewhat crude but effective. Don't make the mistake of putting too many lines on them; estimating between existing lines, even though they are some distance apart, is as effective as trying to read graduations so close together that you need a magnifying glass. Remember that the function of setting circles is simply to bring the object searched for somewhere within the field of the eyepiece. You will bring it to the center of the field by looking at the object itself, not by following the setting circle readings.

To make more precise circles than the ones described you will need a lathe or a dividing engine and the ability to operate machinery within close tolerances. Lacking this, you can have a good machinist do the job for you but it is likely to be an expensive operation. You may, of course, wish to buy your circles from any of the many suppliers of them. If you do, buy the largest circles your pocketbook will permit. A pair of printed plastic 8-inch circles cost about $5–$10.

Using Setting Circles

The calculation of sidereal time need be done only once during any observing period. Determine the sidereal time for the hour when you plan to be at your telescope. Then set your watch or an alarm clock for the sidereal time found, remembering that sidereal time makes use of a 24-hour rather than a 12-hour system. A clock or watch, even though running on solar time, will deviate from sidereal time by only 10 seconds each hour, which is a negligible difference during a short observing session. To set the RA circle for any given celestial body, look up the RA of the object on your star chart, compare it with the sidereal time as shown on your clock, and make the necessary subtraction to find the proper hour angle to set on your RA circle.

For those who find the computations involved in determining sidereal time tedious or difficult to understand, here is a quick and easy method that eliminates most of the mathematics. It is just as accurate as any other method.

The process makes use of the fact that all stars move across the sky at the same rate and a change in hour angle for one star represents the change for all others. You pick out a "clock" star whose motion will serve as a base for all the rest. This should be a bright star, preferably near the celestial equator, that you can easily locate on your star chart, or whose declination and right ascension you can look up in a table. (See appendix XIV)

Set the declination axis of the telescope for the declination of the star. Then move the telescope in hour angle until the star is centered in the field of view. Turn the RA circle on its shaft until its reading coincides with that of the star, and clamp it in place. Now set the hands of an alarm clock at 12 o'clock and start it ticking. The alarm clock is now keeping sidereal time, or close enough to it for all practical purposes.

To find the proper setting for any other star you need only to look up its coordinates on your star chart. Move the telescope in hour angle until the reading on the circle is the RA of the new star, plus whatever the clock reads.

Then adjust the telescope in declination to the proper reading, look in the eyepiece, and you will find the new star close to the center of the field.

TELESCOPE DRIVES

The completed setting circles enable you to find any object within the visible sector of the heavens and to center it in the eyepiece. But once this is done, how can you keep it there? What controls must you have on your telescope? This depends on the particular object you have in the field of your eyepiece. If it is a body outside the solar system—nebula, cluster, double star, etc.—you need only to control the motion of the polar axis, since these objects are fixed in declination. But if it is a member of the solar system you must apply controls to both axes, for the sun, moon, and planets move both in declination and right ascension. The moon, for instance, may vary as much as 14′ in declination in an hour; the sun and planets vary less but still change in declination. The same is true for motion in right ascension. Each member of the solar system varies to a greater or lesser extent from the "fixed" speed of the stars across the sky.

We have already provided for motion in right ascension by equipping the polar axis with a worm gear. If a flexible cable is attached to the shaft of the worm, you can easily control the motion of the telescope. To complete the system of controls, however, you may also want to provide a worm gear for the declination axis. This can be done by using a worm which can be disengaged from the worm gear. The telescope may now be turned freely in declination except when fine control is desired. But, for the most part, no controls are needed on the declination axis unless one wishes to photograph the members of the solar system.

The ultimate in controls is to attach a drive of some sort which will move the telescope in right ascension at the same rate as the stars. This is a fairly easy device to rig up if you wish only to achieve an approximate rate sufficient to keep the object in the field long enough for visitors to have a look. But it is a more difficult problem if you wish to produce an exact rate of motion, which will make it possible for the telescope to be used photographically over long periods of time. Let us consider the latter case first.

The power supply is the first consideration. We must find one which (1) has a very steady rate, and (2) has enough power to turn the telescope and its attachments. Many of the devices used by amateurs fail to satisfy one or the other of these requirements. Small electric motors, alarm-clock works, spring-driven phonograph motors, and similar power sources develop idiosyncrasies which are hard to control. Probably the most efficient source of power is a synchronous electric motor geared to the polar axis. These motors run at a steady rate if the electric power supply which energizes them is controlled at its source. Since the electric clock has become so widely used, electric power companies are careful to regulate their voltage. Synchronous motors are a little more expensive than others but their dependability more than makes up for the extra expense. There are many sources of these motors, ranging from those designed to drive clocks, which are perfectly satisfactory for light telescopes, to the heavier models required for larger telescopes. A $\frac{1}{20}$-hp synchronous motor is rugged enough to drive any telescope, provided the supplementary gear system is lined up properly and friction cut to a minimum. Synchronous motors range in price from $15–$40 and may be obtained in a wide variety of rpm ranges. The most popular for telescope purposes is 1 rpm, and most gear systems are based on this speed as a starting point for the system.

The 1-rpm output of the motor must be transformed to a rate at the polar axis shaft which will turn the telescope in sidereal time. Actually, because of the refraction of light by the atmosphere, the telescope must turn a little more slowly than the sidereal rate. Atmospheric refraction bends the light from a star and makes it appear to lag a little behind its real position, hence the rate at which the telescope must be turned varies for declination and hour angle, since the atmosphere is thicker toward the horizon than it is overhead. But since the rate cannot be varied so that it will be correct for all parts of the heavens, an average rate, which takes into account most of the factors involved, has been worked out. This is 24 seconds a day slower than sidereal time. Since the sidereal day is 23 h 56 m 4 s, the polar axis must be made to turn a complete revolution in 23 h 56 m 28 s, equivalent to 2/2873 or .000696 rpm. Fortunately the transformation from 1 to .000696 rpm is much simpler than the figures given would make it appear to be.

Simple gear systems are not difficult to work out if we remember that the increase or decrease in revolution speed of any pair of gears is always equal to the number of teeth on the driving gear divided by the number of teeth on the driven gear. A shaft revolving at 2400 rpm, fitted with a gear having 10 teeth, will drive another shaft at 240 rpm if the second shaft has a gear with 100 teeth, or,

$$2400 \times \frac{10}{100} = 240$$

The gears ordinarily used in telescope drives are of three types: spur gears, in which one gear meshes with another and the faces of the gears are parallel to each other; change gears, which are similar to spur gears but are arranged so the teeth may be readily unmeshed; worm gears, in which the worm part is a screwlike arrangement on a shaft and engages a worm wheel. Worm gears are usually set up so that a complete revo-

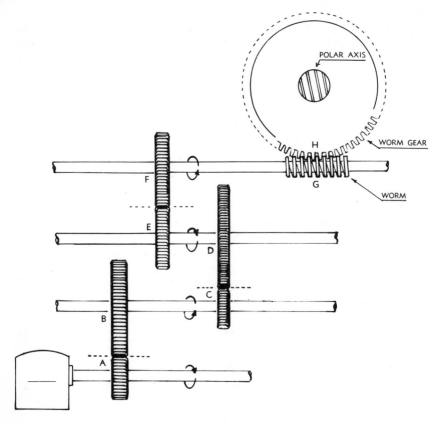

Gear arrangement for a four-shaft drive system.

$$1 \text{ rpm} \times \frac{20}{66} \times \frac{14}{56} \times \frac{34}{37} \times \frac{1}{100} = \frac{2}{2873} \text{ rpm}$$

| Gears | Gears | Gears | Gears |
| A & B | C & D | E & F | G & H |

Another system:

$$1 \text{ rpm} \times \frac{16}{52} \times \frac{40}{68} \times \frac{10}{26} \times \frac{1}{100} = \frac{2}{2873} \text{ rpm}$$

lution of the worm moves the worm gear through the part of a revolution equal to one tooth. The speed of the shaft connected to the worm gear is therefore reduced by a factor equal to the number of its teeth. Worm gears are used for large reductions of speed and are not reversible, while spur and change gears are ordinarily used for smaller reductions and can be reversed. The gears supplied by the Boston Gear Works and similar concerns may be obtained in a wide variety of sizes of all three types. They are made of iron or bronze. The latter are corrosion-resistant, but iron gears are perfectly satisfactory for telescope drives as long as they are well taken care of.

By connecting pairs of gears in a series we may decrease the original shaft speed of the motor to any desired value for the shaft we wish to drive. Depending on the complexity of the system, we may use as many pairs of gears as necessary. But we must be careful, in any gear system, that the design calls for gears which are usually stocked by the manufacturer.

One way to demonstrate how such systems work is to use an example. In the diagram on the next page, the worm gear on the polar axis has 100 teeth. This gear system works as follows to produce the desired reduction:

These two systems require three shafts and bearings for each shaft, in addition to the motor and polar axis shafts. The fewer shafts and bearings required to support the gears, the less expensive the gear system, the less space required, and the simpler the arrangement is to set up. On the other hand, there is a limited number of gear combinations which will produce the final shaft speed desired, 2/2873 rpm; and, when we reduce the possible combinations to a single intermediate shaft between motor and polar shafts, we run into gears of odd tooth count which manufacturers ordinarily do not stock. Unless you have a lathe and the equipment to cut your own gears, you will find it necessary to have them made up as a special order, and this is expensive. You could, for instance, use the following system:

$$1 \text{ rpm} \times \frac{10}{65} \times \frac{1}{221} = \frac{2}{2873} \text{ rpm}$$

You have achieved a very simple gear system here but you must pay for it in the form of a 221-tooth worm wheel, which is hard to find and expensive after you have found it. Another system presenting the same difficulties is as follows: [2]

$$1 \text{ rpm} \times \frac{10}{85} \times \frac{1}{169} = \frac{2}{2873} \text{ rpm}$$

There are also several two-intermediate shaft possibilities. Two which amateurs have found useful, but which require 120 instead of 100 tooth worm gears, are indicated below:

$$1 \text{ rpm} \times \frac{14}{37} \times \frac{17}{77} \times \frac{1}{120} = \frac{2}{2873} \text{ rpm}$$

$$1 \text{ rpm} \times \frac{19}{69} \times \frac{27}{89} \times \frac{1}{120} = \frac{2}{2873} \text{ rpm}$$

[2] Note that in all these systems, the numbers appearing in the denominators are all factors of 2873 or some multiple of it, while those appearing in the numerators are equal multiples of 2. The combinations are limited because of this mathematical requirement.

The diameters of the gears mentioned above can be found easily after we become familiar with a few of the terms used by gear makers. This is by no means an exhaustive list, but it will serve for the basic concepts. *Pitch* indicates the number of teeth per inch of diameter of the gear. Thus a 20-pitch gear whose diameter is 1 inch has 20 teeth, while a 20-pitch gear of 3-inch diameter has 60 teeth. Used in reverse, if the pitch and number of teeth are known, the diameter of the gear may be found. If a 32-pitch gear has 56 teeth, its diameter is $^{56}\!/_{32}$, or 1.75 inches. But a 20-pitch gear with 56 teeth has a diameter of $^{56}\!/_{20}$, or 2.8 inches. The size of gears found in this way is called *pitch diameter* and is twice the distance from the center of the gear to a point approximately at the center of a tooth. This is the point at which the gear teeth bear on each other.

In setting up a gear system, the two difficulties most commonly encountered are (1) finding, after the system is complete, that the final shaft revolves in the wrong direction, and (2) having backlash. Wrong direction of revolution can be taken care of by turning the motor end for end or rewiring the motor connections. The shaft of the polar axis must turn so that the telescope sweeps from east to west. Backlash is the result of play in the gears, usually the final-stage worm gear. It decreases with the size of the worm gear, but it cannot be eliminated completely simply by making this gear larger. It can be reduced greatly by using a braking system on the polar axis. Wrap a band of flexible metal around the axis; secure one end to the base plate of the mounting and secure the other end by means of a spring to maintain tension on the band. If the telescope is well balanced, very little tension will be required.

The systems described are for the sidereal rate of the stars after correction has been made for refraction. They cannot be used to follow the motions of the members of the solar system without supplementary guiding. The moon presents the most serious problem in this respect. But it is possible, by using interchangeable gears, to set up a system which can be used for both sidereal and lunar rates. One such combination

$$1 \text{ rpm} \quad \times \quad \frac{20}{56} \quad \times \quad \left[\frac{28}{66}\right] \quad \times \quad \frac{34}{74} \quad \times \quad \frac{1}{100}$$

produces a rate which is 24 seconds slow on sidereal time. But if the pair of gears shown in the brackets is changed to $^{27}\!/_{66}$, the system now moves the telescope at an average lunar rate. Use change gears for the 28- and 27-tooth gears so that they may be easily substituted for one another.

Simple Drive Mechanisms

If you wish to set up a drive system which will hold a star in the field for short periods of time—say 15 minutes or so—the elaborate systems described above are unnecessary and a simple system may be substituted. Most of the simple systems make use of a falling weight to turn the polar axis, just as the weights in a grandfather's clock turn the hands in accordance with the rate of the pendulum. The key to such systems is to find a means of controlling the rate at which the weight decends, thus controlling the motion of the telescope.

One possibility is to wrap a $\frac{1}{16}$-inch steel cable around a drum placed on the shaft which holds the worm. The cable is placed so that a weight on one end will turn the telescope toward the west, and the retarding force is attached to the other end. This mechanism can take several forms. One popular method is to attach a rack at some point in the cable. The rack engages the driving gear of an alarm clock, and the escapement of the clock forces the weight to descend at a steady rate. The arrangement is not hard to rig up and is simple to operate. Notice that the alarm clock does not provide the power to turn the telescope; it acts only as a brake to control its rate.

Other amateurs have used the piston of a hydraulic cylinder as a brake. The cylinder is provided with a needle valve so that the hydraulic fluid can escape at a steady rate, thus regulating the motion of the piston as the cable pulls against it. The fluid may be water in warm climates, although regular hydraulic oil such as is used in the braking systems of cars is better. One of the new fluids, a derivative of the silicones, is better than either oil or water since its viscosity does not change with temperature and very little adjustment is required.

The point of attachment of the cable may be the worm shaft, as described above, or the cable may be wrapped around the polar axis itself. In this case, sheaves or pulleys must be provided to support the cable and keep it clear of other parts of the telescope. Some amateurs have attached special arms to the polar-axis shaft, supplemented with lever systems in an attempt to provide a more uniform rate of motion.

The important objective in any of the above systems for approximate rates of motion is to keep whatever you use as simple and inexpensive as possible. If you spend time and money on a drive system for your telescope, don't waste it on apparatus which will give you only approximate results. Use a gear system and a synchronous motor and work from the beginning toward a drive which will produce a sidereal rate. On the other hand, if all you want is a device to keep an image in the field for a few minutes, a little ingenuity will go a long way toward working out one that will satisfy your desire without damaging your pocketbook.

CHAPTER 15

The Observatory

YOUR COMPLETED TELESCOPE probably has cost somewhere between $50 and $350, depending on the number of extras that you have included in its construction. If you made use of machined parts in the mounting, installed a drive and setting circles, have a good selection of eyepieces, and added a finder, you may have approached the second figure given. Most amateurs find that they have spent around $125 for their finished instruments. Their reward for the time and energy they have expended is a telescope worth many times this amount.

As a basis of comparison, a professionally made 8-inch reflector, complete with drive and setting circles, costs from $600 upwards. Its approximate equivalent, a 4-inch refractor, starts at $950. Yet few amateurs seem to realize how valuable their instruments are. It is the common fate of portable reflectors to be stored in attics, garages, or cellars, with resulting damage to optical systems and mountings. The damage arises not only from the inadequate storage system but also from constantly moving the instrument from one place to another.

Almost any form of outdoor storage is preferable to this sort of banging around as long as it offers protection to optical surfaces and bearings. Even a crude shelter eliminates the necessity for setting up and dismantling the telescope every time it is used and makes it instantly available. Even if there is some damage from the elements it is usually far less than that occasioned by eager and awkward assistants who simply can't be made to understand that a telescope is a precise and fragile instrument, no matter how sturdy it looks.

PROTECTING THE OPTICS

The first consideration in an outdoor location is protection of the optical surfaces. A tightly fitting dust cap over the open end of the telescope tube is essential; and, if this can be supplemented by a covering for the mirror itself, so much the better. If the tube is of an open framework design, access to the mirror is easy; but in a closed tube of paper, aluminum, or fiberglass, an access hatch must be cut in the lower end of the tube. In fiberglass, this can be done with very little loss of tube strength if the access hatch is made just large enough to permit the insertion of the mirror cover. A 3×9-inch longitudinal cut near the base usually is sufficient. The section cut out can be hinged and replaced on the tube after a strip of fiberglass tape has been added to help seal the edges when the "door" is closed. A loosely fitting cap, lined with soft absorbent material such as a section of quilted mattress pad, gives the mirror all the protection it needs and also serves as a guard against moisture. The cover must not actually come in contact with the mirror surface; and this can, of course, be prevented if the cap is made deep enough. If the cap is brought indoors periodically and thoroughly dried, it will protect the mirror surface for years.

Usually the dust cap on the open end of the telescope is sufficient protection for the diagonal mirror, provided the eyepiece is not removed. Rubber protective caps for eyepieces are stocked by most of the war-surplus companies or, if not available, can be made by gluing a piece of rubber over the end of a section of cardboard tubing whose inside diameter will fit the outside of the eyepiece.

These protective devices should be used under any circumstances, whether the telescope is stored indoors or out, but of course are essential for the latter.

CLEANING OPTICAL SURFACES

In spite of all precautions, your mirror will eventually get dirty. Before you rush at it with a can of scouring powder and a wire brush, remember that a very dirty mirror works better than a poorly cleaned one, and that even the most careful cleaning procedures sometimes cause scratches. But if you feel you *must* clean your mirror, the following procedure is perhaps as safe as any.

Remove the mirror from its cell and gently brush its surface with a camel's hair brush, the kind photographers use on their camera lenses. Wet a freshly laundered piece of old toweling—the older the better, since filler material and dust have been washed out by repeated launderings— in distilled water. Pour a solution of detergent and distilled water on the mirror and allow it to stand for a few minutes. This will soak off the sharp edges of organic material and loosen embedded silica particles. Dump off the solution, add more, and *gently* swab the surface of the mirror. Don't apply pressure, tilt the mirror, hold the cloth by one corner, or swab with a circular motion. Rinse several times with distilled water—under no circumstances use tap water since this usually contains dissolved minerals which will react with the aluminum to produce a cloudy surface. If the mirror is really clean, there will be no clinging droplets when you stand it on edge to dry. If there are one or two droplets, remove them with the corner of a piece of clean white blotting paper without touching the mirror's surface.

PERMANENT MOUNTINGS

Any permanent outdoor installation of the telescope makes it possible to use a pier instead of a tripod to support the mounting. Many amateurs, as we have mentioned, use a 4-inch water pipe set in a concrete base—if the telescope is located in cold-weather areas this *must* extend below the frost line—or build the whole pier from concrete. If the area around the pier is paved with concrete or other material, space should be left between the pier and the inside edge of the paving material. Otherwise any vibration set up in the pavement is transferred to the pier and thence to the telescope. It is a good idea to elevate the pavement an inch or so above ground level to ensure good drainage, but if it is set too high people stepping over the edge in the darkness can get rather nasty falls. Make the pavement at least ten feet on a side; a few extra cubic feet of concrete costs little in money or effort and a larger area around the telescope prevents visitors from crowding the observing space.

Not everyone is fortunate enough to have a nearby outdoor observing area available, of course. A roof top is possible as an alternative, although here it is troublesome to provide a suitable pier which will be vibration-free. This can be partially eliminated by providing rubber pads placed at several points around the base of the pier or at the ends of the tripod legs.

SIMPLE HOUSINGS

The easiest and least expensive way to protect the telescope from the weather is to make a covering of canvas or plastic whose open end can be tied securely around the base. This provides good protection against everything except very high winds or the curious passer-by.

A roll-on housing is perhaps the least expensive means of protection from each of these hazards. You can usually pick up lengths of light T-iron from junkyards or wrecking companies. These make excellent double tracks for ball-bearing grooved wheels. To ensure easy operation, the wheels should be at least 4 inches in diameter. Ball-bearing wheels of this size can be obtained at many hardware stores, and only four are needed for a housing of moderate size and weight. Make the track by using the T-iron with the shank of the "T" upward. If it is placed in a concrete pavement, sink the track below the level of the pavement; although this will make snow clearance more of a problem, the track will not be a hazard on a dark night. The housing itself can be of any available material; make it out of marine plywood, siding, or canvas primed and sealed and tacked over a wooden framework. Build it only large enough to accommodate the telescope. Because of its open-base construction, it cannot be used for storage effectively; and building it larger than necessary serves no useful purpose. The lightest and strongest material that you can find for a housing of this kind is the polyethylene plastic used for greenhouses. It is very tough and will last practically forever.

The door at the end of the housing must be made large enough to provide plenty of room for the telescope as the housing slides over it. When the housing is in place over the telescope, padlock it in position at both ends so it cannot be pushed or tipped off its track.

SLIDING-ROOF OBSERVATORIES

The housing described above protects the telescope from the weather but does nothing for the observer. The breezes which spring up on a cold winter's night are not conducive to observing, but their chilling effects can be reduced by some sort of windscreen so that the hardy stargazer is at least moderately comfortable. Four walls and a sliding roof overhead give this sort of protection and constitute the most inexpensive observatory possible to build. But here a word of caution is in order. If you have reached the stage where your interest in astronomy includes building an observatory, build for the future and not for the present. The chances are good that your 8-inch telescope may sooner or later seem inadequate and you will want to replace it with a larger instrument. This is the typical pattern for the amateur. Having built his own telescope, he is as much interested in the instrument itself as in what he can use it for, and this leads him on to ever larger and more complex models. A building which will house a 6- or 8-inch telescope will not be adequate for a 12-inch, so it is better to build for the larger instrument in the first place. Even though the observatory is to be a simple one, size is of paramount importance. Any observatory less than 12 feet square will sooner or later become too small for your purposes. The same is true for the pier which supports the telescope; place your 8-inch telescope on a pier made for one twice as large and you will not regret it later.

In any observatory, the pier and the building must be two separate entities. Otherwise vibrations arising from the building are transmitted to the telescope. So build the pier first, then place the building around it.

Two views of a sliding roof observatory, built by George T. Keene for his 12-inch f/4.3 telescope. The dimensions of the building are 10 by 10 by 4 feet. The roof slides toward the north to give an unobstructed view of the southern skies. (Photos by George T. Keene)

This is true only for an observatory of this type; for a circular one you will reverse the process, for reasons explained later. For solidity and permanence concrete has no substitute as a building material in the pier. Excavate below the frost level down to solid rock or gravel, make a form to place over the hole, add a few reinforcing rods whose upper ends are threaded to hold the base plate of the mounting, and fill with concrete. Even at the present price of concrete mix, you cannot build a pier from any other material at lower costs. If you plan to have a concrete floor in the observatory, be sure to leave a "breathing space" between pier and the inner edge of the floor to prevent the vibrations noted above.

Good design of a sliding-roof observatory. (Photo by William Phillips)

The walls of the observatory serve only to support the sliding roof and to keep out wind and weather. Unlike those of a house, they need not be built to retain heat. In fact, it is better if they do not. From the discussion of heat effects on mirrors in an earlier chapter, you can see why it is undesirable to heat an observatory. It is very much to your interest to see that inside and outside temperatures in a small observatory remain as nearly the same as possible, otherwise the air currents produce fatal local conditions of turbulence as far as good "seeing" is concerned. Two-by-four studs covered with weatherproof wallboard are as good as stone walls from a practical, if not an esthetic, viewpoint. They should be painted with a light reflective coating on the outside of the observatory and with a dark color on the inside. This is to make sure that any heat trapped in the building during the day will be quickly radiated away as soon as the slide is opened in the evening.

The height of the walls is a matter for your own judgment. If they are too high, the pier of the telescope must be built up, otherwise the section of the sky near the horizon will be obscured. But as the pier is increased in height, the farther the eyepiece will be from the ground and the higher the stepladder you must have to reach it. On the other hand, if the walls are too low, your tall guests will be in constant danger of fracturing their skulls when the roof is in place. You won't, yourself, after a few experiences, but the result may be a permanent stoop from moving around the observatory in a crouching position. A good compromise is to make the walls about 6 feet high. This gives room enough for a door, yet requires a stepladder only a few feet high for you to reach the eyepiece comfortably, even when the telescope is pointed toward the zenith.

The roof may be made in one section or two. Each has its advantages. If it is in one section, you must add a fairly long track (12 to 15 feet) on one side of the building so that it may be slid clear of the telescope. T-iron for a track and 4-inch grooved wheels for rollers are best for this purpose because such an arrangement is self-clearing of ice and snow. Needless to say, the supporting structure for the roof runway must be adequately braced because a 12-foot-square roof, even if made of the lightest possible materials, will still weigh a considerable amount. Again as in the simple housing previously described, the polyethylene plastic used for greenhouse covering is the best possible material with which to cover the framing of the roof. This particular plastic is no more expensive than ordinary roof coverings, is amazingly strong, and will last for years. If laid over a framework of well-braced 2×4's, it will support two feet of heavy snow without sagging or breaking.

The single-section roof is not difficult to make or to operate. The pitch should be made as steep as possible, the only limitation being how much of the sky it shuts out, and of course it should be pitched toward the telescope when the roof is open because in this way it will obscure a minimum amount of the heavens.

A roof made in two sections is preferred by most people. It is twice as easy to handle since each section weighs only half as much and need be pushed only half as far. This may sound like a specious argument since in the long run the same total weight must be pushed the same distance to open up the observatory completely. The point is, of course, that it is much easier to handle two light sections of roof, one after the other, than a single heavy section in one great effort. But the greatest advantage of the double-section roof is that it gives greater protection in adverse weather conditions. If the wind is blowing, only half the room need be opened to view a particular section of the sky.

At the same time, if the roof is to have an adequate pitch, the view toward the lateral sides of the observatory—that is, the sides toward which the halves of the roof travel—will be interfered with considerably and much of the sky near the horizon will be lost to view. The remedy for this must be to lengthen the track for each section so that the sections may be pushed farther away from the building. This of course increases the labor involved in opening and closing the building but it is a relatively small price to pay for a simple and workable system. Most people do not consider this a difficulty, although there is another which must be taken into account. This is the problem of attaining a weather-tight seal between the two roof sections when the building is closed. Since the joint occurs directly over the telescope, great care must be taken to see that the sections overlap just enough to prevent leakage and yet do not interfere with opening and closing the roof, especially when the roof is covered with snow. The final advantage of the two-section sliding roof is that the supporting structures are placed on opposite sides of the building and present a pleasingly symmetrical appearance whether the room is open or closed.

In any observatory where sliding roofs are used, hooks must be placed at strategic positions so the sections may be fixed in any desired position. This is particularly important when the building is closed. Five or six hundred pounds of roof sliding off on inquisitive small boys is no laughing matter.

DOMED OBSERVATORY

Most amateurs tend to view building a dome as something beyond their powers because of the curved surfaces involved. There is little doubt that they are greater than those for a sliding-roof observatory, yet the fact is that no special skills and a minimum of equipment are required. As verification of this view, the observatory at the Millbrook School in New York was constructed by schoolboys working with a teacher who knew little more than they did, and an electric drill and a band saw were the most complicated equipment used. This building turned out to be a circular structure of 17-foot diameter with a rotating dome made from a silo top. The base of the building was constructed of cinder block, and the silo top was obtained from a nearby manufacturer at a discount. But these materials need not have been used. Many other amateurs are in the process of building highly satisfactory observatories from other materials.

The base can be constructed of wood in a variety of forms. Perhaps the most utilitarian is a rectangular building with a dome at one end, thus providing space for a work room at the other end. If the building is divided by a partition, the work room can be heated as long as it is well insulated from the dome end. Other workers have constructed a square building. Both types of base for the dome have the disadvantage of attempting to support a round dome on a square base, since the base ring of the dome must be strongly supported at all points. A more popular and sturdy base is a structure with many sides, six to ten of them. These are harder to build, but the extra cost and labor are well worth it.

Whatever form the building takes, make it larger than you think you need.

Telescope makers are like boat owners; they are never satisfied with what they have. A structure big enough for your original 8-inch telescope will hardly provide space for a 16-inch scope.

In general, a circular building for an observatory is more pleasing to the eye —perhaps this is what we are accustomed to from photographs of the big observatories. As a compromise among the many forms we may choose from, let us consider a circular building of modest proportions—say 12 to 15 feet inside diameter—as a beginning point. Later, if you wish, you can add storage and work space by extending the door area. We choose this form of structure since a circular wall offers perfect support for the dome rotating above it, and also because it is no more difficult to build a circular wall than a plane one. Our choice would be to use cinder block because it is light, easy to handle, and relatively inexpensive. Like the walls of any building, these must have adequate footings to prevent the sagging and cracking which result from uneven settling of the structure. Dig a circular ditch and fill it with concrete until it is perfectly level all the way around.

How high should we build the wall? Unless you wish to take the chance of your guests fracturing their skulls on the lintel of the door, build the wall 6 feet high. This will involve raising the pier of the telescope or building up the floor. We chose the latter, and added two steps just inside the door for easy access. A hatchway, hinged on one side, was put in place to prevent people from falling into the stairwell when the telescope was in use.

Capping the Wall

At this point a band saw is the most useful piece of equipment you can use because you must saw out enough curved sections of ¾-inch plywood to make three complete circles, double laminated for strength. If your observatory is 14 feet in diameter, the distance around the wall will be 44 feet, or 11 sections, each 4 feet long and 6 inches wide. Make a pattern by using a stout cord with a loop at each end. The distance between the ends of the loops must be exactly 7 feet. Then tie a knot 6 inches from the end of one of the loops. You now have a compass in which the total length of the string will give you the curvature of the outside of the curved sections, and the distance from the knot to the far end of the loop will be right for the inside curve. Inscribe these curves across the short dimension of a 4 × 8-foot sheet of ¾-inch plywood. Cut this piece out and use it for a pattern for the remaining pieces you will need, first checking to see if it fits the top of the wall. All told, you will need 66 such pieces. Why so many? You need three complete rings, one for the top of the wall, one for the base of the dome, and a third which you will cut into two semicircles for the sides of the aperture in the dome. Trim the ends of each piece so that when placed together they will make a smooth, regular circle. Lay these out on a flat surface, then glue another layer on top of the first, laying the center of each so it covers the joints in the sections below.

Place the completed ring on top of the wall, and secure it in place with bolts paced every few feet. To make sure the bolts are secure, fill some of the

This 17-foot observatory was built by a group of 14- to 17-year-old boys with no outside help. The base is cinder block, surmounted by a converted silo top with hemispheric slide. It holds a 12-inch telescope.

openings in the top layer of cinder blocks with concrete, and while it is still wet insert the bolts, using round-head bolts 4 inches long and ⅜ inch thick. Now check the ring to be sure it is perfectly level.

While the concrete is drying, start work on the second and third rings. You will have plenty of time for this operation since concrete needs at least two or three days to cure properly.

Once the rings are finished, you must decide what to use as rollers so the dome will turn easily. Again there seem to be almost as many systems as there are builders. But one basic rule must be observed. Don't use too many rollers under the impression that the more there are, the easier the dome will turn. Quite the contrary is true, for each extra roller adds more friction to the system. Four rollers are enough to support even the heaviest and largest dome, provided they are at least 6 inches in diameter. If possible, use ball-bearing rollers, either grooved or with flat surfaces; little wheels are an abomination. But it must be said that there are exceptions to these statements. A teacher and his students in Beebe, Arkansas, used 120 golf balls rolling on a grooved track. They say their dome turns with the pressure of only one finger!

If wheels with flat rims are used, they may be attached to the wall plate so the dome plate runs over them, or vice versa. The former is probably the better system of the two. But if either method is used, a third wheel, placed horizontally, must be added to prevent lateral movement of the dome. These wheels bear against the skirt of the dome, an addition we mention below. Still another method is to have grooved wheels running on a track made of aluminum cast in an upside-down T form. Sections of the aluminum can be bent to form a perfect circle in a machine shop, or you can do it yourself by hammering it against a heavy wooden form cut to the curvature you wish. If you decide on this system, attach the rollers to the building plate and the track to the bottom of the dome plate. Such an arrangement has two advantages; it prevents any lateral motion of the dome, and it also provides a means of turning the dome without muscle power. If the shaft of one of the wheels is extended, a V-belt system can be attached so the dome can be turned by an electric motor. Wires and a switch can be mounted on the pedestal of the telescope and the observer can turn the dome without leaving his position at the eyepiece. But unless the drive wheel and the track are in firm contact with one another, mechanical difficulties can arise. One way to prevent this is to mount a weight immediately above the wheel, thus providing good contact. After the dome ring has been assembled and the wheel system placed in position, lift the dome ring into position and make any adjustments necessary to see that the dome ring moves easily and smoothly.

Most domes weigh several hundred pounds, and most people build them on the ground and lift them into position when finished. But this takes manpower, and there is always the possibility of damage during the operation. To avoid these complications, it is just as easy to build them in place.

Constructing the Dome

After the dome plate is in place, the next step is to position the side pieces which determine the dome aperture. These have already been built; they are the semicircles mentioned earlier. A common mistake is to place them too close together. Place them so their distance apart is at least one-quarter of the dome diameter. This gives an observer inside the building a wide view of the heavens and also cuts down on the number of times he must move the dome to follow an object as it moves across the sky. Make sure the main ribs on each side of the aperture are perpendicular to the dome ring; then brace them securely with temporary bracing. The next step is a tedious one—making and mounting the ribs. These need not be as strong as the ribs you used for the dome aperture. They can be made from laminated plywood: ½-inch strips 4

An attractive 8-foot observatory built of plywood and ⅛-inch tempered Masonite. It contains a sturdily mounted 8-inch f/6 equatorial Newtonian telescope. (Observatory and photo by Charles Wierzbichi)

inches wide, glued and nailed together, and having the same curvature as the base ring. Depending upon the size of the dome you will need about nine of them, their butts spaced at equal intervals around the dome ring. Many workers, tired of cutting curves, use fewer ribs. But this creates a form midway between a rhomboid and a hemisphere. If you use bracing between the ribs they should be curved according to their placement. Braces forming a circle near the top of the dome must have sharper curves than those near the base plate.

The Dome Covering

The dome covering must conform to several requirements. It must be light, strong enough to support winter snow and ice, and above all flexible enough to adapt to curved surfaces. Plywood satisfies the first two, but is almost impossible to bend to the form required. It is possible to make ⅜-inch plywood more flexible by cutting kerfs on one side of it. But this is a tricky business because if the kerfs are cut too deep the smooth side of the plywood has a tendency to crack. Nevertheless, a dome made this way is strong and permanent.

Sheet aluminum satisfies all three of the criteria, and so does tempered Masonite and thin fiberglass sheet. In each case panels, called gores, must be cut to fit the opening between each pair of ribs. The gores must be cut very carefully, the edges of one fitting tightly against the edge of its neighbor. Finally, the joined edges should be covered with waterproof tape to seal out the weather.

Another very successful dome covering is the heavy plastic used to cover greenhouses. Careful tailoring of this material will result in a gore that covers several ribs at a time. Before deciding on this type of covering, talk to a greenhouse owner. He will probably tell you that it provides a watertight, permanent, and easy-to-apply covering.

Other Dome Possibilities

Silo Tops

The use of a silo top for an observatory dome presents many interesting possibilities. A silo top comes in sections which are easily and permanently fitted together to produce a strong and rigid, yet exceedingly light, dome. The normal opening at the top of the silo is just the right size for the zenith opening and is reinforced with a ring for greater rigidity. The slit must be cut from the side of the top, which is a disadvantage, but this can be done with a pair of rugged tin snips, supplemented by a hacksaw for cutting through some of the built-in ribs. The edges of the slit thus produced must be strengthened with angle iron, bent to conform to the curve of the silo and bolted securely in place. You must also cut a base plate for the dome just as with other forms, even though the silo top is manufactured with a ½-inch or larger steel rod which runs completely around the base. But this rod provides an excellent means of attaching the dome to the base plate, and the result is increased strength in this vital area.

There are many advantages in adapting silo tops as observatory domes. They are extremely light—a 17-foot dome made from one weighs only 250 pounds —they never need attention because the aluminum produces a self-protective coating of aluminum oxide which lasts for years, and they need little or no interior bracing. Finally, they cost as little, completely assembled, as any other form of observatory dome.

Molded Fiberglass

One particularly fine example of the molded fiberglass dome comes from D. Ratledge of Adlington-Chorley, England. He made a mold from fine-mesh chicken wire, lumber, and plaster and used it to make four quarter-hemispheric segments. These, when fitted together, made a ribless hemispheric dome. Each segment has three identical sides and three right-angle corners. Instead of using wooden ribs, he added a 4-inch rib to each side of each quarter-hemisphere. When bolted together it was an extremely solid, tough dome in which the bottom rib of each segment formed the dome plate. Finally, he cut the dome opening after everything was assembled.

What makes the Ratledge observatory unique is that he also constructed the base from fiberglass. Using another homemade mold, he cast curved panels which, when bolted together, formed the base of the observatory, 9 feet in diameter. But the base does not support the dome. Instead, he placed four posts at equal intervals inside the walls, and on these he mounted a roller system,

Two views of Mr. Ratledge's fiberglass observatory. (Photos by David Ratledge)

Support system for Mr. Ratledge's observatory dome. (Photo by David Ratledge)

as the photographs show. The whole project is the result of imagination and industry, and the end product is a beautiful and useful building.

Either the fiberglass dome or its companion, the adapted silo top, has an advantage possessed by no other type. It may be used as a planetarium as well as observatory dome, a feature which makes it particularly desirable to those who wish to use an observatory as a source of instruction in astronomy. Even one of the cheap globe type of planetarium projectors, set up inside the dome, will project a reasonably accurate facsimile of the heavens upon the curved roof. The shift from the representation of the constellations inside the observatory to the actual observation of them outside, or to the viewing of key parts of constellations through the telescope, is quickly and easily made and is an invaluable aid to teaching. If the observatory is large enough and if the inside of the dome is painted a flat black, the inside representation of the stars is almost as fascinating to the viewer as the outside actuality.

The Dome Slide

There are three types of slide commonly used in small observatories. The single-unit slide which moves sideways to expose the slit, the double-unit slide which splits down the middle to distribute weight when opened, and the hemispheric slide which slips over the top of the observatory on rollers are all popular with amateurs.

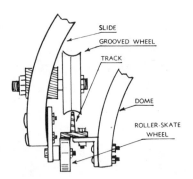

Detail of track and roller system for dome slide.

For a very light dome, the first type is perhaps the most awkward to build and use since it shifts a considerable proportion of the total dome weight to one side of the center of gravity, producing balancing difficulties. Moreover, its edge, when fully extended, acts as a wind scoop. Unless the whole slide is well anchored against its supports, this can be very bothersome in windy weather.

The second type, or double-unit slide, maintains balance much better but has the disadvantage of a seam down the middle which is hard to close against the weather. The third, or hemispheric type, has the greatest deficiency of all because, by its nature (since it must cover more than half the

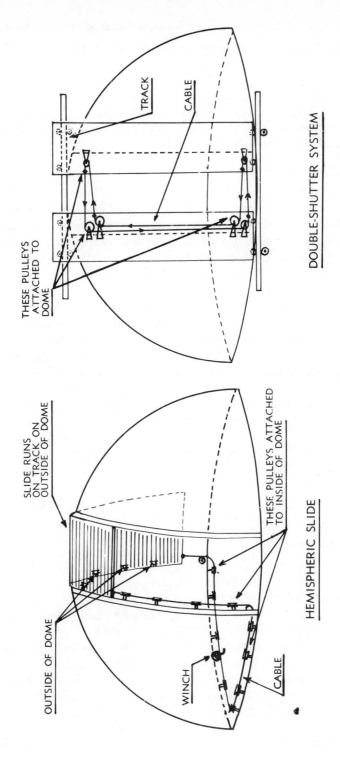

TRACK

CABLE

DOUBLE-SHUTTER SYSTEM

THESE PULLEYS
ATTACHED TO
DOME

SLIDE RUNS
ON TRACK ON
OUTSIDE OF DOME

THESE PULLEYS ATTACHED
TO INSIDE OF DOME

HEMISPHERIC SLIDE

OUTSIDE OF DOME

WINCH

CABLE

Pulley systems for dome slides.

dome when closed), it shuts out the zenith area of the heavens. This can be overcome by hinging a section at the lower end which can be flipped back when the zenith is to be observed. But here again a wind problem is created and, unless the hinged section can be firmly secured, the observer is likely to have some uncomfortable moments on windy nights. All other things being equal, the double-unit slide is to be preferred not only for the sake of convenience but for appearance as well.

Double-Unit Dome Slide

Making a double-unit slide is not a difficult task. You must provide straight tracks at the bottom and top of the dome for the halves of the slide to travel on, and each slide section must have its own rollers. A pulley system (see diagram) should be rigged to open and close the slide from within the observatory. The rest is easy, since it is as simple to build a curved slide as a flat one. The surface of the slide follows that of the dome when the slide is closed, and the only difficult part of the construction is making the curved girders which form the outside edges of each slide section. These may be wood, made in the same way as the ribs previously described (page 260), or aluminum or steel T-girders bent to shape. If the latter, the metal must be heavy enough not to sag against the dome. Crosspieces placed about a foot apart connect the girders, then a covering of plastic is stretched over the framework thus provided.

Use 2-inch grooved wheels attached to the reinforced ends of each slide for the sideways travel. Attach flat-surfaced wheels, such as those from roller skates, to bear against the under surface of the tracks to prevent the other wheels from slipping off the tracks. Finally, you must provide a framework for the outer edge of each slide which will bear lightly against the dome to effect a good closure against the weather when the dome is closed, and of course there must be an overlap of the meeting edges of the slide sections for the same purpose.

The Hemispheric Slide

The hemispheric slide is preferred by many because it produces a tighter closure of the dome slit than any of the other types. If the dome is a perfect hemisphere, this slide can be made in one-piece semicircular form; unfortunately many domes sag enough to cause binding or sticking, especially after a heavy snowfall. Hence it is best to make this kind of slide in jointed sections, each 3 or 4 feet long. The sections are connected to one another to form a continuous covering over the dome slit, and each section is supported by small wheels running in a pair of U-shaped tracks bent over the top of the dome, one on each side of the slit. The bottom section of the slide—that is, the section which will be on top of the dome when the slide is opened—must be hinged and free to open to make the zenith visible. otherwise 15° or 20° of the heavens will be lost to view.

A pulley system and a small winch such as those provided for boat trailers is necessary to raise and lower this type of slide. These pulleys are placed on the inside of the dome and on the outside to serve as a track for the light steel cable which controls the slide movement. Turning the handle of the winch in one direction raises the slide and in the opposite direction, lowers it (see diagram).

The hemispheric slide is more difficult to build than either of the other two types, but once it is operating smoothly it gives less trouble, especially under snow or icing conditions. This is because the tracks on the slit are concealed under the slide when the observatory is closed. Gently raising and lowering the slide for a very short distance is usually sufficient to get rid of even a heavy snowfall. A light metal covering such as aluminum sheet, or the plastic material recommended earlier, is ideal for this type of slide.

FURNISHING THE OBSERVATORY

There are several pieces of equipment which are essential to the smooth operation of any observatory. While it is impossible to anticipate the wants and needs of any individual observer, at least a partial list of items can be made which all observers find useful.

General Equipment

1. A sturdy stepladder, preferably placed on rollers so that it may easily be moved about. It is most helpful to your observing comfort if you build a sliding seat on the stepladder which you can move to the most convenient position. A rack placed in an accessible position to hold flashlight, pencils, books, etc. is also a great help.

2. A desk the top of which is at least 16×24 inches (the size of the largest charts you will use). This should preferably be built far enough from the floor so you can use it while in a standing position. The added height also permits the addition of extra drawers for eyepieces and other equipment.

3. A well-shaded red light for the desk. Red light is the least damaging illumination to night vision. Use the cylinder type of low-wattage bulbs instead of the ordinary round ones. Cover the bulb with red plastic and place a shade so that the light is directed on the chart and not around the observatory. Any light in an observatory must be used sparingly and kept as dim as possible.

4. A flashlight whose lens is covered with red paint or plastic. This is a great boon when visitors are present so that you can point out stepladder positions, the location of the eyepiece, etc., without impairing their vision.

5. Some folding chairs. Helps in keeping visitors fixed in position.

6. A small short-wave radio which will pick up time signals from WWV or CBU (see table of frequencies in the appendix). This is invaluable for time exposures because you can listen to the time instead of consulting your watch. This is, of course, a luxury item.
7. A sidereal clock. Another luxury item. Buy one when you have everything else you need; it will impress your visitors. In the meantime, use an ordinary alarm clock (a necessity) for sidereal time.

Charts and Books

There are dozens of books, charts, and other publications which are useful in an observatory. The following is a minimum list, each item of which is worth its weight in gold in a small observatory (see appendix for a more complete list).

1. A set of monthly star charts culled from the back pages of *Sky and Telescope* magazine. These are the clearest guides to the constellations it is possible to find and are invaluable for quick reference. Mount one for each month in a cardboard folder.
2. Deluxe or regular edition of the Skalnate Pleso *Atlas of the Heavens*. Perhaps the finest star charts available. Included with a set of the charts is a transparent grid for a quick determination of star coordinates.
3. *Norton's Star Atlas*. Excellent star charts plus descriptive lists of interesting objects which can be seen with small telescopes. Also contains much useful data on the solar system.
4. *Field Book of the Skies* by William Olcott and R. Newton and Margaret Mayall. This is a splendid revision of a standard work. It is packed with information about every aspect of stargazing, whether with telescope or naked eye.

CHAPTER 16

Using Your Telescope

That first look through a newly completed telescope is usually exciting and stimulating. The moon and planets look great, as do the larger star clusters, the great gaseous nebulae, and the nearby galaxies. You take great pride in demonstrating the virtues of the instrument to your neighbors and to anyone else who shows interest. Then, if you are like most enthusiasts you try to pick up some of the fainter objects in the sky, like close double stars, planetary nebulae, and the more obscure and distant galaxies. After that, one of two things happens. You decide that the 8-inch telescope you have built with such loving care isn't quite big enough and that you would like to try something larger, or you begin to lose interest.

But you don't have to follow this pattern. If you can find some purpose in looking around the heavens, you may find yourself wrapped up in a hobby that will last for years. Perhaps you begin to suspect you aren't using the instrument to its fullest capabilities. Or perhaps your observing techniques can be improved to the point where you discover you haven't exhausted your enthusiasm in exploring the heavens. Here are a few possibilities that may help.

Have you set your telescope up away from street lights or the glare of passing automobiles? Do you try to observe in bright moonlight? Do you try to find objects in the vicinity of Jupiter, Venus, or other bright objects? Are you using a bright flashlight to illuminate your observing area? Any of these factors can obscure what you see through your eyepiece. You can improve the clarity of your observations by taking a few simple precautions. Wait for dark, moonless nights. Ban all flashlights unless the lens is covered with red plastic, the darker the better. Be sure your eyes are dark-adapted by waiting a half hour in a dark

269

place before using the telescope. True, there isn't much you can do about bright outside lights except to change your observing area, but if you pay attention to the other factors you will find that the view through your eyepiece is very much improved.

Try to choose nights when the stars don't twinkle. Twinkling stars indicate that there is turbulence in the atmosphere. The effect is magnified by your telescope, often to the point where the "seeing" is impossibly distorted. Look for those evenings when, to the naked eye, the sky seems to have great depth. This means that dust and other particles in the atmosphere are at a minimum, and on such nights celestial objects stand out sharp and bright. Moonless evenings when the stars don't twinkle and the sky is transparent are rare; don't waste them.

Check your telescope constantly. Be sure it remains in collimation. Use dew-caps on both primary and secondary mirrors to keep them free of moisture and dust. Also, in using your eyepieces be careful about the placement of your eye. High-powered eyepieces have very short eye relief. To use them effectively, the eye must be close to the eye lens. The reverse is true of course for long-focus low-powered eyepieces. Here it is often useful to use a rubber eye-cap to keep out extraneous light and to center the pupil on the Ramsden disk. Very dim objects can be made to appear sharper by using averted vision. In other words, don't look directly at the image, look to one side of it. Above all, don't stare into the eyepiece.

Finally, make an effort to acquaint yourself with the heavens. Learn the constellations, and the names and Greek letter designations of some of the bright stars they contain. Actually, the Greek alphabet isn't hard to learn since it's so similar to the English equivalent. So study the skies without benefit of telescope. Then, when you want to find an object in a particular constellation, you will know where to point the telescope.

PROJECTS FOR AMATEUR TELESCOPES

The surest way of maintaining an interest in astronomy is to take part in one of the many projects open to amateur observers. Most owners of small telescopes feel that their observations are of interest only to them or to other observers. In fact, quite the opposite is true. There are several professional organizations that are eager to have the contributions of amateurs, and that find them extremely valuable. We list some of them below.

The ALPO Program

The Association of Lunar and Planetary Observers gives amateurs the opportunity of making contributions to the body of information about the solar system under the guidance of experts. It is a common impression that the photographic satellites have shown all we need to know about the planets and their moons, but this is far from the truth. There is still much to be learned about our neighbors in space, but it can be done only by constant observation. The great telescopes of the world are occupied with other projects—the Voyager Project is no longer functional because of lack of funds—so dedicated amateurs are left to take up the slack.

ALPO has been helping interested amateurs—novices, advanced observers, even professionals—since 1947. It has a membership of more than 750, who pay a very nominal fee for the privilege. Each member receives the quarterly journal, called *The Strolling Astronomer,* but the benefits go far beyond the magazine subscription. It provides, to quote its brochure, "a service for the novice who wishes to develop his knowledge and observing technique, the advanced amateur who wishes to contribute to solar system astronomy, and for the professional who can participate in group studies and systematic patrols." In addition, there is a convention each summer for exhibits, business sessions, and the presentation of papers.

Because any consideration of the solar system involves a very large body of information, the ALPO is divided into sections for specific areas of study. Each section is headed by a Recorder, who keeps track of the data provided by the members, and who also supplies observing kits and information applicable to his area of study.

The Training Program

This program deals with fundamental skills such as how to record observations. It also emphasizes training the eye, estimating the clarity of the atmosphere, and rating "seeing" conditions. Since many observations require drawings, the program instructs in the development of drawing skills and the acquisition of drawing materials.

The Sections

The Recorders for the various sections are spread all over the United States from Massachusetts to California and from Texas to Wisconsin. You may obtain their names and addresses, as well as membership in the association, by writing to Prof. Walter H. Haas, Box 3AZ, University Park, Las Cruces, N. M. 88003.

The sections cover almost every aspect of the solar system. They include the following:
1. The Lunar Section, subdivided into three subsections; the Lunar Incognita Program, the Lunar Transient Phenomena Program, and the Selected Area Program.
2. Six planetary sections, each with its own Recorder. They include Mercury, Venus, Mars, Jupiter, Saturn, and a Remote Planets Program for Uranus, Neptune, and Pluto.
3. A Minor Planets section to study the asteroids. This section publishes its own journal, *The Minor Planets Bulletin.*

Comet Kohotek. Fifteen-minute exposure on Tri-X film. (Photo by Steve
Reed)

4. A Comets section, of great interest now because of the passage of Halley's
   Comet.
5. Lunar Eclipse study; not a section by itself since lunar eclipses are visible
   only about once a year from any given locality. Nevertheless they are of
   high interest when they do occur.

Variable Star Observations

Another program in which the contributions of amateurs are of great value
concerns variable stars. Most people think of the stars as points of light of
unvarying brightness. But many stars vary in brightness over short or long
periods of time. Of the more than 25,000 variables at least fifteen groups have
been identified. The simplest case is when two stars revolve about one another,
alternately obscuring and augmenting the total light coming from the combination. But most variables fall into groups named for the constellation in which
the first one was discovered. Hence there are variables named for Cepheus (the
Cepheids), Lyra (the RR Lyrae stars), and so on, each with different characteristics and periods. For example, the Cepheids are young stars found in the arms
of spiral nebulae and are divided into two groups. Classical Cepheids have
periods of 1.5 to 28 days; long-period Cepheids take more than 28 days to

Great Nebula in Orion (M42, M43). Eleven-minute exposure on 103 af film. (Photo by W. E. Hamler)

complete their cycle. On the other hand, the RR Lyrae variables are old stars that occur in globular clusters and have periods measured in hours. Space does not permit a discussion of the many other types of variables, but such information is available from an organization called the American Association of Variable Star Observers. The AAVSO is located at 187 Concord Avenue, Cambridge, Mass. 02138. Membership is open to anyone who has a real interest in this fascinating hobby.

Observing the Variable Stars

Why the interest in variable stars? Largely because the behavior of some of them is unpredictable. Some shine brightly, then drop to near invisibility. Some seem to explode (the Novae); many are irregular in their variations. One class (the Cepheids) serve as measuring sticks for the dimension of the universe. All can be observed with accuracy by amateur astronomers.

What do you need to observe them? First, of course, is a telescope in good working order, and a knowledge of how to use it. Setting circles are a great help in locating variables, although the charts supplied by the AAVSO show star patterns in the vicinity. A good eye for determining relative brightness is essential, but once again the charts give some help in the form of nearby stars whose magnitudes have been measured right down to the nearest tenth. Magnitude has nothing to do with the dimensions of a star; it is simply a measure of its apparent brightness. The brightest stars are of zero magnitude; those of fifth magnitude are 100 times less bright. Each magnitude in between is less bright by the factor 2.512; the dimmest star that can be seen with the naked eye is sixth magnitude. Beyond that, telescopes must be used. Some of the variables are fifteenth magnitude or less. At the other end of the scale, objects brighter than zero magnitude are indicated by negative numbers. Thus Venus at her brightest is −4.4 and the sun is −27.9.

Timing the observation of a variable is less critical than one might think; it is measured in tenths of a day. But the notation used is confusing to those who are jut beginning to study them; it is based on the Julian Day calendar. Each day that has passed since Jan. 1, 4713 B.C., has been numbered consecutively. Since that date, nearly 2.5 million days have elapsed. For example, Sept. 2, 1982, is Julian day 2,445,215. But this huge number doesn't present any problem. Calendars have been printed showing the Julian day as well as the days of the year. The Nautical Almanac also contains a table of Julian days. In practice, you convert the time you make the observation into Universal Time, then change this time into the appropriate tenth of a day. Suppose, for example, you make an observation at 1:30 A.M., EST, on Sept. 2, 1982. This is 0630 UT, or 0.3 days. So you write the following notation on your record sheet: 2,445,215 3. To save time and space you would omit the first three digits since they don't change, and you also omit the decimal point.

All the information we have given here, and much more, can be obtained from the AAVSO. We include a sample chart here so you can see what kind of information will be supplied.

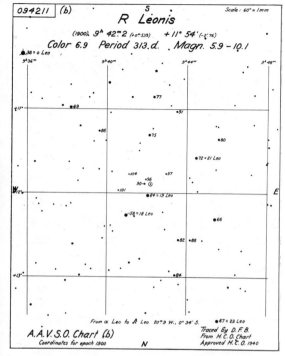

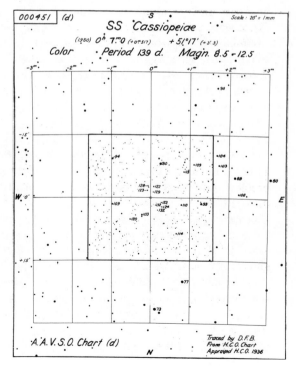

Two examples of AAVSO charts.

Observing the Meteors

In its ceaseless travel around the sun the earth constantly passes through the left-over debris of comets. Most of it is made up of tiny particles about the size of a pinhead or a grain of sand, but some are rock-sized or larger. They don't become visible until they strike our atmosphere. Then their impact on air molecules produces heat from the friction of their passage, and this in turn heats the particles until they glow. The tiny ones actually burn up, but the larger chunks may pass on through and hit the earth. The particles are called meteoroids; if they are big enough to get through they are called meteorites. The streaks of light we see in the sky seem to be very close, but they are from 60 to 90 miles above us.

If the earth passes through a dense layer of debris we see anywhere from 5 to 50 meteors per hour, all seeming to come from the same point in the sky called the radiant, and this is called a meteor shower. Occasionally the number of meteors in a shower reaches a very high number. For example, a shower originating in the constellation Leo in 1966 streaked into the sky at the rate of 10,000 per hour. These came from the debris left by the great comet of 1886. Since the earth passes through the leavings of any comet in the earth's orbit, we should expect to see the resulting shower at the same time each year, and this is what happens. But not all meteors have their origin in comets. On any dark night as many as a dozen may be seen that are not associated with any known meteor shower. They are referred to as *sporadic* meteors.

Meteor showers are named for the constellation in which their radiant lies. Thus there are Lyrids, Geminids, Perseids, etc. Some are named for the comets in which they originate, such as the Giacobinids. Radiants are spread all over the heavens, appearing in at least 25 different constellations. Although they appear at about the same time every year, the duration of different showers varies greatly, from half a day for the Quadrantids to 25 days for the Saggitarids.

Recording Meteors

Before you think of recording meteor observations, plan your meteor watch so you do it under the best possible conditions. One of these, unfortunately, is the time of night. The best results are obtained after midnight. The earth moves around the sun at a speed of 18.5 mph; meteors moving more slowly than that never catch up with it. Early in the evening you are moving away from the meteor stream, late at night you move into it. You can observe a striking example of this phenomenon by watching the Perseids, a meteor shower that occurs each year about the middle of August. Before midnight you may see about 25 meteors per hour; after midnight the number increases to a maximum of 50 or more. Not only this, but the latecomers are particularly spectacular. They are bright yellow. Some of them appear to explode; many more leave trails of smoke.

Unless you don't mind having a stiff neck after a night of meteor watching, don't stand up. Instead take a sleeping bag and spread it over a reclining lawn chair. Have your recording pad near at hand and use a red filter on your flashlight. If you plan to time the meteors, bring along a shortwave radio and tune it to the time ticks from station WWV. Have a star chart handy and estimate the length of the meteor trail by noting stars along its path.

If you are going to keep records, do a good job, or don't bother. What should you record? The brightness of the meteor, its color, the length of its path, its duration (usually 2 or 3 seconds, but you can get an accurate estimate by counting the ticks from WWV), the time at which it occurs, and anything unusual about it such as apparent explosions, any noise it makes, changes in color, smoke trails, etc.

If, after a few experiences, you wish to pursue the matter further, get in touch with the American Meteor Society, Department of Physics and Astronomy, State University of New York, Geneseo, N.Y. 14454.

You can expect to see meteor showers in any month during the year. Space does not permit listing all of them, but here are a few that are particularly worth watching:

Shower	RA	Dec.	Date of Maximum	Number per Hour
Quadrantids	15h 28m	+50°	Jan 3	45
Aquarids	22h 30m	−2°	May 4	20
Draconids	15h 12m	+58°	June 28	50
Perseids	2h 4m	+58°	Aug 12	50
Orionids	6h 20m	+15°	Oct 20	25
Geminids	7h 28m	+32°	Dec 13	50

Observing Lunar Occultations

The moon, in its eastward passage across the sky each night, passes in front of a number of stars. Such eclipses are called occultations and are of great interest to astronomers. Two organizations in particular supply information to interested amateur observers. They are The United States Naval Observatory (Occultation Timing Division, Washington, D.C. 20390), which issues predictions based on the geographical location of the observer, and *Sky and Telescope* magazine (49-50-51 Bay State Road, Cambridge, Mass. 02138), which will supply a booklet containing the major occultations for each year.

In observing an occultation, you will be looking for just one bit of information: the exact instant of time when the star disappears behind the edge (limb) of the moon. This sounds simple, but actually it involves a good deal of planning and meticulous attention to detail.

First, of course, is your telescope. It must be rock-steady since any vibration from wind or other causes can spoil the accuracy of your observation. You must use high magnification both to cut off some of the glare of the moon and to catch the exact moment when the star disappears. You must also determine the exact latitude and longitude of your observing station and its height above sea level. Careful study and measurement of the geodetic survey map of your area will give you this information. You can determine your location within 10 to 20 minutes of arc and your elevation to 3 feet. The maps can be obtained from stores which specialize in engineering and geological equipment; if you can't find such a store, write to the United States Geological Survey Commission, Denver, Col. 80225. Each map costs $2.00. Even if you decide not to participate in occultation studies, the maps are valuable for other purposes.

Timing Occultations

Here you need equipment you may not already have; a shortwave radio capable of picking up the WWV time signals, a good tape recorder, and an accurate stopwatch that reads to within a tenth of a second. Equipment of this nature was widely used by those who participated in the search for early satellites in the MOONWATCH program in the late 1950's and is easy to use. The WWV signals come in as ticks at 1-second intervals; the stopwatch supplies the intervals between ticks.

Finally, you need a good set of detailed star charts so you can identify the stars whose occultations you will measure.

As pointed out earlier, this project requires acute attention to detail. Because it also requires some expensive equipment, it should not be entered into lightly. It is for serious observers who are intent on making a real contribution to this particular phase of astronomy.

SOME OTHER PROJECTS

So far we have listed activities in which the amateur can make a contribution to general astronomical information. Now we consider some others for your own entertainment and information.

Sunspot Observation

Before we go into the details of this topic, a warning is in order. The sun must never be observed directly through a telescope unless the instrument is equipped with protective devices—Herschel wedges, filters of proven safety, masking devices that permit only a tiny fraction of sunlight to enter the telescope, and the like. *Direct observation of the sun can damage or destroy your eyesight.*

But the sun can be observed safely if its image is projected onto a screen placed outside the eyepiece. An arrangement such as this is not hard to rig up simply by attaching a right-angle clamp to the telescope tube with a large piece of cardboard placed in the jaws of the clamp. The farther the cardboard is from the eyepiece, the larger the image and the easier to observe it. Once a sharp focus is achieved, sunspots will show up clearly and in detail. Then use a soft pencil to trace the outline of each sunspot. It helps greatly if you draw a circle on the cardboard beforehand and center the sun on the circle. If you replace the cardboard every day the sun shines, you will soon have a complete record

The Trifid Nebula (M20). Fifteen-minute exposure on 103 aE film. (Photo by John Sanford)

of the way the spots seem to move across the face of the solar disk and of the changes in them from day to day. The apparent movement of the sunspots is due to the rotation of the sun, but the spots also have a motion of their own. They usually appear about 40° above or below the solar equator, then gradually move toward it. Most spots last only a few days or weeks, but some linger as long as 27 days, which is the sun's period of rotation at the equator. Like the planet Jupiter, it rotates faster at the equator than at the poles.

Photography of the Heavens

With some relatively simple additions to your telescope, you can make a photographic record of the skies, starting with star trails, constellations, and the Milky Way, and ending with pictures of the elusive planet Mars and the fascinating Messier objects. All that is required is a 35-mm camera and the means of attaching it to your telescope. For some objects, you don't even need the telescope. For more information on the details of celestial photography, see the next chapter.

An open cluster in Scutum (M11). Fifteen-minute exposure on 103 aE
film. (Photo by John Sanford)

Hunting the Messier Objects

Tracking down all the objects that make up the Messier Catalogue is perhaps one of the most interesting and instructive hobbies an amateur can
pursue. These fascinating shapes and forms are spread all over the heavens and
most are located so they can be best observed from the Northern Hemisphere.

The French astronomer Charles Messier was primarily interested in comets.
In his careful study of the skies during the latter part of the eighteenth century
he observed many fuzzy objects that were obviously not comets because they
remained fixed in position. To avoid wasting his time if he happened upon the
same objects again, he plotted the position of each one and included them all
in a list. Other astronomers, notably Pierre Mechain, added to the list until
today it stands at 109, if a few uncertainties are included. A pictorial list of
the Messier objects can be found in a book by Mallas and Kreimer, and
descriptions of all of them are in the author's *Telescope Handbook and Star
Atlas*.

On star charts the Messier objects are designated by the prefix M plus a

The Omega nebula (M17). Twenty-five-minute exposure on 103 af film. (Photo by Steve Reed)

number. They occur in a variety of shapes, sizes, brightness, and nature. Some are spherical balls of as many as 100,000 stars, called globular clusters. Other groups, in which the stars are more widely spread out and which contain from 50 to 300 stars, are known as open clusters. Then there are diffuse clouds of gas, the diffuse nebulae, and small well-defined gas clouds called planetary nebulae. Finally there are innumerable galaxies of all sizes, shapes, and descriptions.

Some of the items in the Messier list are visible to the naked eye as fuzzy patches of light, but binoculars or telescopes are needed to define them into recognizable objects. Care must be taken in all cases to use powers that show them to best effect. The Beehive cluster in Cancer, for example, can be separated into individual stars in a black, moonless sky by persons with extremely sharp vision, but binoculars or low power brings out its sparkling beauty. High power yields only a few widely separated stars in the eyepiece field. High power will also magnify some objects into invisibility, as in the case of the spiral galaxy in the constellation Triangulum. On the other hand, the beautiful globular cluster in Hercules shows up best under high power.

Your 8-inch telescope will reveal all of the Messier objects in all their beauty even better than Messier saw them. If you are like most amateurs you won't want to stop after you have seen them all. The thousands of NGC objects still await your attention. Then there are the double and multiple star systems in all their glittering colors. Separating what appears to be a single star into its components tests not only your skill as a telescope maker but also your ability as an observer.

ASTRONOMY CLUBS

Most people find that sharing their interest in astronomy more than doubles their enjoyment and adds a new dimension to the fascination of exploring the heavens. This is perhaps why there are so many clubs spread all over the country. The advantages are manifold: exchanging information about telescope making, having the opportunity to use the telescopes others have made, talking about techniques of observing and building, and using observatories and clubhouses. If you live anywhere near such a group it may be well worth your while to consider joining with them. Try going to one of their star parties. You will probably find that the exchange of information about what you see is a fascinating experience.

If there is not such an organization in your vicinity, you may want to form your own group. No matter where you live there are sure to be others who will be happy to share information with you. Once you have located such a group, write to other astronomy clubs to find out what their experiences have been. *Sky and Telescope* magazine periodocally lists amateur clubs and gives information about their ativities. You can also find specific directions about forming a club from magazines and books. One of the best is P. Clay Sherrod's *Complete Manual of Amateur Astronomy.*

CHAPTER 17

Celestial Photography

I N chapter 16 we mentioned that a very satisfactory use for your fine new telescope could be in making a photographic record of the heavens. In this chapter we must point out some of the difficulties, as well as the satisfactions, you may encounter if you succumb to the astrophotography bug.

Let's start with equipment. The best camera for this purpose is a 35-mm SLR (single lens reflex), although others may serve. The SLR is light, versatile, and is receptive to the addition of other equipment. It can be attached to some convenient point on the telescope tube or directly to the eyepiece holder. Most SLR cameras have bulb and infinity settings and locking cable releases. Their shutters, or whatever other means are used to expose the film, work smoothly and without vibration. They also usually come equipped with fast lenses. Some, however, may have additional features (electronic shutter controls, etc.) that may actually be a hindrance to their use in astrophotography.

Celestial photography has less in common with the terrestrial type than one might think. Its subject matter—stars, planets, galaxies, nebulae, etc.—is in constant motion, even if very slow compared to moving objects on earth. Their emitted or reflected light is dim and needs fast cameras and fast film to register properly. Other light in the night sky is a source of constant interference in its variations due to atmospheric turbulence, sky transparency, street lights, moonlight, and other heavenly bodies such as planets. Even variations in latitude influence the amount of light entering the camera lens.

In spite of these difficulties, the amateur who wants to photograph the heavens will find it a very satisfying experience provided he approaches it in simple, easily accomplished steps. It need not be expensive; beginning requires only a minimum of equipment. Previous experience with terrestrial photography is desirable but not essential.

STILL PHOTOGRAPHY

You can have a lot of fun photographing the trails the stars leave on your film as they move across the sky. You don't even need your telescope for this. Your camera, a tripod or other solid support, and some film are the essentials. Startrail photographs like the one illustrated are simply time exposures. These must be long exposures if you want the star trails to be of good length. The rate of a star across the sky is only 1° in 4 minutes. You can get some idea of this scale if you consider that the angular distance between the pointers in the Big Dipper is 5°. A star, moving through a comparable distance, would take 20 minutes. To obtain star trails covering a quarter of the sky requires 3 hours, long enough to fog your film if there is much extraneous light in the sky.

For this type of photography use films such as Plus-X or others of low ASA ratings. Many of the "fast" films (Tri-X, etc.) are subject to what is called

Can you recognize the constellation Orion in this star-trail photograph? (Photo by Howard Snow)

reciprocity failure—the tendency for a latent image to decay almost as fast as it is formed after a short period of time. If the time period is only 3 minutes, for example, there is little point in exposing the film for periods longer than that.

So for star-trail photography, wait for the black, moonless nights and use slow film. You will not only get some good photos, but you will also discover things about the stars that perhaps you didn't know. For example, try taking a picture of the North Star with the circumpolar constellations circling around it. You will obtain proof that the North Star is actually not the north point of the heavens but is located about a degree away.

CONSTELLATIONS WITH STILL PHOTOGRAPHY

After you have mastered the technique of photographing star trails, try "stopping" the star motion by using short exposures. With fast film, a reasonably fast lens in your camera, and modern developing techniques, you can use still photography to make your own catalogue of the heavens from a fixed tripod. Films such as Royal-X Pan whose normal ASA rating is 800 may be force-developed in Phenidone type of developers like Microphen or Clayton P-60 to ratings as high as 2000. When these are used with fast lenses—f/3.5 or better—exposure time may be as low as 2 or 3 seconds. Exposures of this length will not completely "stop" the stars, but the blurring produced will be almost imperceptible.

When you use this technique, one of your primary concerns will be the size of field covered by your camera. But there is a simple formula which will take care of the problem. It is:

$$W = \frac{57.3 \times S}{f}$$

where W = width of field in degrees
     S = film size
     f = focal length in inches

Let's suppose you have a 35-mm camera with a lens of focal length 105 mm. The film used is 24 × 36 mm. What area of the sky will it cover? According to the formula, the 24-mm dimension will cover

$$W = \frac{57.3 \times 24}{105} = 13.0°$$

and for the 36-mm dimension,

$$W = \frac{57.3 \times 36}{105} = 19.0°$$

This is a respectable chunk of the sky, and such a camera would give you good photographs of the constellations, of satellite passages or meteor trails and of the aurora. On the other hand, the images formed on the emulsion would be small and this camera would not be a very good one for photographing the moon or planets.

GUIDED-CAMERA PHOTOGRAPHY

The combination of fast films, fast lenses, and superdevelopment is often not available to the amateur. Nor are they necessary, since increasing the exposure time will produce the same or better results. But longer exposures require that the camera be moved at the same rate as the stars. This isn't hard to accomplish because your telescope mounting is ready-made for the purpose. You will have to change the counterbalancing a little to take care of the added weight, after which you mount the camera on the telescope tube and photograph the stars directly, not through the telescope. If the telescope has no drive system, hand driving by turning the slow-motion knob of the cable drive on your polar axis at a speed to follow the stars across the sky will be necessary. But you will find this tiring and difficult, and a smooth-working drive system is almost essential for any exposure longer than a few seconds. The drive system takes all the work out of the operation and you will use the slow-motion knob only to smooth out any inaccuracies in the driving rate. In this operation the only function of the telescope optical system is to guide the camera; it plays no part in the photography itself.

There are two or three ways to make this kind of photography easier and more pleasant. Mount the camera out of the way but within easy reach of your position at the eyepiece of the telescope. This makes the use of a long cable release unnecessary for the operation of the camera, as well as saves you much legwork in running around the telescope to make adjustments. The best position for the camera bracket is as near the tube saddle as you can get it because this is the place of least vibration. But you must, of course, mount the camera close enough to the front of the tube so that you will not photograph the telescope as well as the stars. Make the mounting bracket as simple as possible. A bar such as the ones used with most camera photoflash attachments can be attached permanently to the tube, and the camera can be screwed onto this. Once the camera is set in place and lined up with the optical axis of the telescope, start the drive mechanism and allow it to run for a minute or two to take up any slack in the gears. Then open the shutter of the camera, taking care not to jar the telescope.

Exposure Time for Guided-Camera Photography

The exposure time varies with the speed of the emulsion, the speed of the lens, and the brightness of the object to be photographed. The general rule is to make it as short as possible since this cuts down on the effect of atmospheric turbulence and lessens the strain of guiding the telescope. Tri-X is satisfactory if exposed with a moderate-speed lens (f/2.5) for up to 10 minutes. Develop in D-19 or a similar developer for 8 minutes. The exposure time for any given film varies with the square of the f-number used. For example, if the correct exposure for an f/2 lens is 1 minute, it must be 4 minutes for an f/4. The exposure time should not be prolonged in the hope of pulling in faint stars in the background of a constellation; the usual result of this practice is only to fog the film from sky background light.

If you are fortunate enough to possess more than one camera, use each according to the star field you wish to photograph. As you can see from the formula given previously, using a camera of longer focal length cuts down on the size of the area you can capture on your film. If you have, let's say, a 35-mm camera of 50-mm focal length, you could use it for photographing the Northern Cross, which sprawls over a considerable area of the sky—more than 20°. But if your objective is the Pleiades, your 150-mm folding camera will cover the 1½° area of the constellation, will give you a good plate scale, and, as a result, will need less enlarging.

PHOTOGRAPHY AT THE FOCAL PLANE

As we mentioned earlier, the images actually produced on the film emulsion of a camera of moderate focal length are very small. Because the emulsion is granular in nature there is a certain amount of scattering of light within the emulsion itself; the image of a faint star is "blown up" beyond its actual size and always *appears* larger than the point of light which produced it. Thus a star appears as a *disk* on a photographic plate. It would be convenient if other objects such as the planets and moon were similarly self-enlarging, but this is not the case. We can change the formula we have been using, to get more accurate information of the size of images we may expect from a camera of moderate focal length. The formula now becomes

$$S \text{ (size of image)} = \frac{W \times f}{57.3}$$

In using this formula, be sure that the units match. If the focal length is given in inches, the image size will also be in inches. Similarly, the constant must be changed to fit the units used for angular size of the object, as follows:

for degrees, use 57.3
for minutes of arc, use 3438
for seconds of arc, use 206,265

$$\text{Moon} = \frac{31' \times 200}{3438} = 1 \text{ mm}$$

$$\text{Mars} = \frac{22'' \times 200}{206,265} = .021 \text{ mm}$$

The Horsehead Nebula: a prize-winning photograph by a talented amateur. Thirteen-minute exposure on 103 af film. (Photo by W. E. Hamler)

Uranus with two of its moons. (Photo by John Sanford)

Without going into the resolution [1] of films, which is really the determining factor here, it is obvious that an image size such as Mars is beyond the limitations of any conceivable enlargement, no matter how fine the grain of the film.

The only escape from the dilemma is to use the telescope itself as a camera. Remove the eyepiece from the telescope. Then unscrew the lens from the camera and place the remaining frame so that the film plane falls at the focus of the telescope. The result is a camera of 56-inch focal length because we are now using the telescope mirror as the camera lens. The camera itself has become only the plate holder for the telescope. This will make a vast amount of difference in the size of the moon's image at the focal plane. If we substitute this new camera focal length in the formula that we have just used, the moon now has a diameter of nearly half an inch on the emulsion! Unfortunately, even the increase in image size obtained this way will not help us with the planets, but later we find a way to overcome this problem too.

[1] The resolution of a film, like that of a lens or mirror, is the power to separate detail in the image. It depends on the grain size of emulsion. A fine-grained film will separate lines drawn as close as 2500 to the inch, or .0004 inch apart.

Another way to use the telescope as a camera is to remove the diagonal from the telescope and substitute a small plate holder and shutter in its place. But this involves radical changes in the telescope itself, and we are looking for a way to adapt it to photography without altering its structure.

So we leave the diagonal alone and use a camera frame as a film or plate holder simply by mounting it outside the adapter tube (the outside tube of the eyepiece holder). If the camera is of the reflex type, so much the better, because we can use its own built-in system for focusing. But if it depends on an auxiliary focusing arrangement which cannot be used here because it will not line up with the telescope optics, we shall have to adapt the camera slightly. This can be done by opening the back of the camera and holding a piece of ground glass at the image plane to obtain a sharp focus before we insert the film. If this is too much trouble, we can make an auxiliary plate holder and ground-glass focusing device. A sliding shutter can be constructed to pass across the tube which connects the box to the adapter tube.

From all points of view, however, it is much simpler to use an inexpensive single-lens reflex camera. This is the cheapest and most efficient way of transforming the telescope for photographic purposes because the reflex camera, after you remove its lens, has all three elements that you need: shutter, focusing arrangement, and film holder. Most supply firms for astronomy materials and equipment, as well as several commercial telescope makers, can supply adapter tubes to fit your eyepiece holder at one end and the camera ring at the other. If such tubes are not available for your particular needs, attach a bracket near the eyepiece holder. Insert a rod in the bracket upon which the camera can slide for focusing purposes.

Well, now you have changed your telescope into a camera with a long focal length and you can use it to photograph the moon, nebulae, star clusters, galaxies, and the sun. It may be used for exposures from 1/1000 second to several hours, depending on the object to be photographed and the film used. The extremely bright moon, for example, requires an exposure of only 1/250 second with a fast film like Tri-X. The moon near the quarter needs longer exposures, depending on how much of its surface is illuminated. It is difficult to make any rules about exposure times in this case because the problem is complicated by the difference in light near the terminator (the edge of the shadow) and the illuminated portion. If you expose correctly for one, you underexpose or overexpose for the other. One-fiftieth second is a fairly safe bet, but the chances are that you will waste a little film before you get the exposure time right for your particular telescope.

Photographing the sun is beset with so many dangers and difficulties you may want to leave it alone completely or at least postpone it until you have gained experience with other objects. With an 8-inch telescope,

A star field and the North American Nebula. (Photo by Steve Reed)

no filter placed at the eye of the telescope is adequate because the concentrated rays of the focused sun are dangerously intense. The only safe way is to mask out most of the mirror by placing a piece of cardboard in front of it. Cut a 2-inch hole at one side of the cardboard. This changes your mirror to one of 2-inch aperture, with a consequent loss of resolving power, but it also reduces the illumination from the sun by some 94 per cent. Even so, a filter must be used at the eyepiece. A better solution is to replace the diagonal with a Herschel wedge, a prism which reflects most of the sun's rays off to one side and permits only a very small percentage to reach the eyepiece. This, used in conjunction with a neutral filter placed over the adapter tube, makes it possible to photograph sunspots. Exposures range from 1/200 to 1/500 of a second with a fast film.

Guiding for Prime-Focus Photography

Guiding the telescope during exposures for lunar or solar photography is not a problem because of the shortness of the exposure. Exposure time may be increased to as much as 1/20 second before the telescope drive need be started. The moon's travel across the sky during this brief interval will not appreciably blur your negative. For exposures longer than this, however, the drive system must be in operation.

For objects fainter than the moon, guiding the telescope is the main problem to be solved. Most of your poor negatives will owe their defects to errors in keeping the telescope pointed at the object to be photographed. Exposure time for nebulae and galaxies will range from 5 minutes to an hour; and, even if your sidereal drive is very accurate, other factors such as atmospheric refraction effects will change the position of the image on your emulsion, with resultant loss in sharpness. So you must use the slow-motion knob on your polar axis to a considerable extent during the exposure. Be careful that you don't become too fussy here, for overguiding can produce as poor results as underguiding. Most drive systems have defects which operate in one direction—the star image has a tendency to drift along a specified line—so you must bring it back periodically, the time interval being dependent on the amount of drift. If you overcorrect, you double the original error.

You are now using the optics of the telescope for photographic purposes, and it is difficult to use them for guiding at the same time. This *can* be done by using a guiding eyepiece, offset to one side of the photographic field, which is centered on a star image provided by the telescope optics. Such a device is hard to rig up; and, since our objective is to keep all apparatus as simple as possible, we suggest that you use your finder for guiding purposes. To do this you must adapt the finder to match the conditions under which the photograph is being taken, that is, you must increase its effective focal length as nearly as possible to the focal length of the telescope. Using a Barlow lens (see page 159) and increasing the power of the finder eyepiece is one way to accomplish your purpose. Such an arrangement, for the majority of finders, will not produce an image of any sharpness, but it will suffice for the job at hand. In fact, it is better if the guide star is out of focus because it is easier to see, has a better diameter to keep on the cross hairs, and is easier on the eyesight than a sharply focused object. The guide

Sunspot groups taken at prime focus, using filter and mirror mask.

M-8, the Lagoon Nebula in Sagittarius. Ten-minute exposure on
103a-F, developed 6 minutes in D-19. Prime focus. (Photo by G. T.
Keene)

star need not be the object photographed. Find a nearby bright star and
use that as a guide. It must be kept on the cross hairs at all times, a
tedious and tiring process if the exposure is very long.

The general procedure for prime-focus photography is the same in
principle as that of the guided camera. The drive motor must be started
before the exposure is made. When the driving mechanism has settled
down to its normal rate, open the shutter cautiously. Needless to say,
here is where rigidity in the telescope mounting is of paramount importance, because vibration of any kind is fatal to photography.

Focusing

Focusing the telescope for dim objects is difficult at best, but it can be made
easier if a magnifier is used to be sure the image on the ground glass is as sharp
as possible. You will inevitably ruin some of your negatives through focusing
errors. Indeed, mistakes in focusing rank next to those in guiding as a cause
of wasted film. In the arrangement described here, focusing is done by moving
the camera along its supporting rod. This is admittedly a crude device for the
purpose; even so, you can obtain very good focus if you are careful enough.
The alternative is to attach the camera directly to your draw tube and use the

M-51, the Whirlpool Nebula in Canes Venatici. Fourteen-minute
exposure on 103a-F, developed 6 minutes in D-19. Prime focus.
(Photo by G. T. Keene)

eyepiece focusing device to adjust the camera distance. This can be done if you
are willing to make, or have made, a special fitting as described earlier (page
290) as a coupling for camera and draw tube. One end of the fitting must screw
into the threads which ordinarily hold the lens element, and the other slide into
the eyepiece draw tube. Since the camera opening is usually not the same size
as the standard 1¼-inch eyepiece, this must be a reduction fitting. If you have
a machinist friend or have a lathe yourself, such a coupling device is not hard
to make.

FILMS FOR ASTROPHOTOGRAPHY

In general the same films used for ordinary photography are useful for
heavenly objects. In recent years many new films have been added to the list,
but most of the old ones are still acceptable. Even though the beginner in
astrophotography now has a wide range of choices, it is best to limit selection
to a few. With increasing knowledge it is always possible to experiment with
special-purpose films, or even to transform ordinary film into a high-speed
product, as we shall see below. Before we discuss this topic, however, here is
a list of the films most used by astrophotographers:

Black and White Slower Film	Black and White Fast Film
Plus-X	Tri-X
Verichrome Pan	Ilford HP 4
Ilford Pan F	Ilford XP 1
Ilford FP4	103a-O (blue-senstive)
Variable Speed Agfa Pan	103a-E (red-sensitive)
Kodak 2415	103a-F (panchromatic)

Slow Color Film		*Fast Color Film*
	for transparencies	
Ektachrome 64		Fujichrome 400
Ektachrome 200		Ektachrome 400
	for prints	
Kodacolor 24		Kodacolor 400
Kodacolor 64		Fujicolor 400

SUPERSENSITIZING METHODS

The beautiful computer-enhanced photographs created in the laboratories of the Jet Propulsion organization and others are, at least for the moment, well beyond the grasp of amateurs. Yet there are methods by which amateurs can increase film sensitivity and speed and produce photographs of great quality and beauty. For the most part they are of interest to the more dedicated and experienced amateur astronomers because they require complex equipment that is difficult to use. Nevertheless there is little doubt that with the improvement in amateur techniques and skills these new methods will be widely used in the future. We give a brief description of some of them here.

Cold Cameras

The principle behind cold-film photography has been known for many years, so it can hardly be called a "new" idea. But only recently have cold cameras become popular with amateur astrophotographers. When it became widely known that film chilled to very low temperatures ($-40°$F. to $-60°$ F.) increased in speed and sensitivity, the demand for the technology involved became great. Cooling the film also seemed to reduce grain and to enhance contrast, especially in films sensitive to color. But black and white films such as Tri-X also were affected.

In most chemical reactions, reducing temperature decreases speed, but in film emulsions the opposite takes place. Actually there is some slowing of the almost instantaneous action of light on the chemicals in the emulsion; the gain takes place in the lessening of the decay of the latent image (reciprocity).

Hercules cluster. Fifteen-minute exposure on 103 aE film. (Photo by John Sanford)

The difficulty in the process lies in the means by which the film is chilled. The temperatures required are far below those of water ice, so the only alternative is to use frozen carbon dioxide (dry ice), which has a temperature of $-109°$ F. This is really too cold, but enough heat works into the camera so the actual chill is within an acceptable range. Now the problem is to overcome the other difficulties involved in the process. Dry ice, if handled without gloves, can freeze fingers very quickly. Its effect on film is to make it brittle, stiff, and easily cracked or broken. Consequently only short pieces of film can be used. In order for the dry ice to act on the film, another chamber must be added to the camera, and this adds another difficulty. The proximity of the dry ice to the film cools the air around it, creating a frost which is deposited on the film. Cold-camera enthusiasts have solved the problem in three ways: exhausting the air between dry ice and film by means of a vacuum pump, drying the air so there is no moisture to condense, or filling the space with a block of plastic. The third choice seems to be most widely accepted and to a greater or lesser degree has solved the problem.

The final difficulty, and to most amateurs the most serious one, is the variety of physical problems that are inherent in the process. The camera must be loaded with its short pieces of film and its charge of dry ice (done, of course, in the dark). Then the camera is attached to the telescope, the film exposed and the telescope guided through the exposure time. The camera is removed, returned to the darkroom, the film is brought to room temperature and developed, and the result—a single picture—is examined. It is small wonder that those who are enthusiastic cold-camera buffs often work in pairs, one to take the photographs and the other to load the camera.

But there is no doubt that cold-camera results, especially with Tri-X, Ektachrome, and Kodacolor, are spectacular to the point where they are well worth the trouble.

Hypersensitizing Film with Forming Gas

Forming gas is a mixture of 8 per cent hydrogen and 92 per cent nitrogen. A 35-liter tankful costs about $50 and refills cost half that amount. This may seem expensive, but when films are "soaked" in it under moderate pressure, good things happen, especially to color films and to black and white Kodak 2415. With Ektachrome 400, for example, speed is increased four or five times and colors become more natural. The effect on blue light is enhanced in photographing reflection nebulae, and reds are striking in their realism. Technical Pan 2415, ordinarily just a good film of no great speed but excellent contrast and fine grain, becomes a super film after exposure to forming gas. Speed is increased up to ten times and resolution improves to the point where it is as good as the telescope itself. The film, although black and white, is very sensitive to color variations, especially in the range of red frequencies. Blues are about the same as those of the 103a series of films. Photographers have to be very cautious in their use of 2415 for it records the tiniest imperfections and good guiding is of paramount importance.

There are few places where you can buy hypersensitized film, but you can make it yourself with the proper equipment. If you're mechanically minded and have access to power tools, you can make the equipment yourself. If not, it can be purchased fairly reasonably. You need a pressurized tank big enough to hold a roll of film on an open spindle, a vacuum pump (a hand pump is good enough), a supply of forming gas, a thermometer, and a small thermostatically controlled electric heater. Completely assembled units cost from $200 to $250, but these are one-time expenses for the units last for years.

The process itself is simple, a welcome contrast to the difficulties of cold cameras. Place the film in the pressure tank, first having transferred it from its original package to an open reel. The process will also work if the film is left on its original spindle, but it takes much longer. Close the pressure tank, hook up your vacuum pump to remove the air in the tank, then open the valve to the forming gas supply. When a pressure of 15 pounds per square inch (15 psi) is reached, close the valve, adjust the heater to the desired temperature, set the thermostat, and wait. This whole process is so new that experimenters

The Dumbbell Nebula (M27). Twelve-minute exposure on 2415 film hypersensitized in forming gas. (Photo by W. E. Hamler)

still debate the time the film should "soak" in the forming gas, so you may have to do some experimenting yourself. However, the following conditions seem to be effective. For color film such as Ektachrome 400, try a pressure of 15 psi and a temperature of 25° C. (77° F.) for 5 to 6 days. For Kodak 2415, the same pressure at 50° C. (122° F.) for 4 days seems to work well. Actually, there is only one danger in the forming gas process: If you leave the film in the gas too long, a discoloration called "hyperfogging" results. But at least one worker reports he has left 2415 in the gas for as long as 9 days without damage.

After hypersensitization, rewind the film on its original spindle and place it in a freezer until used. Color film should be used as soon as possible after the hypering process, but 2415 and other fine-grained black and white films may be kept under refrigeration for weeks or even months without adverse effects.

How does forming gas hypersensitizing compare with that of a cold camera? From the point of view of trouble-free operation, forming gas comes out far ahead. But for the quality of results, many photographers who have used both methods seem to favor the cold-camera technique. Some films are improved by one method and unaffected by the other. For example Tri-X is not helped by hypering and is vastly improved in a cold camera, while just the opposite is true of 2415.

Other Methods

There are, of course, other means of increasing film speed. One of them is by extended development, or push-processing. Adding more time in the developer will bring out faint details and increase contrast, it's true, but it also can blacken heavily exposed areas until they become unrecognizable. At the same time it increases grain and images become less sharp. Consequently this means of increasing film speed should be used with great caution, if at all. But there is a way of speeding up the 103a films, remarkable in its simplicity. Suspend the film cassette above a container of very cold water—at least 40° F.—and pull the film down into the water for 2 or 3 minutes. It's a good idea to add a wetting agent such as Kodak Photoflo, otherwise some areas of the film may not become evenly wetted. Be sure to handle the film only by its edges to avoid fingerprints. Hang the film up to dry. When you are sure that it contains no moisture, rewind it into the cassette. Film treated this way becomes noticeably faster but does not remain so for more than a day or so.

Finally, you can use a special developer for certain films. One called MWP-2 (Mount Wilson-Palomar mix #2) doubles the speed of the 103a films as well as Tri-X and HP-4. In the long run, however, force-development is for the professional or the highly skilled amateur.

Saturn. Two-second exposure on Panatomic-X, developed 12 minutes in Phenidone. Eyepiece projection. (Photo by G. T. Keene)

Jupiter, showing the great Red Spot. One-second exposure on Panatomic-X, developed 12 minutes in Phenidone. Eyepiece projection. (Photo by G. T. Keene)

PHOTOGRAPHY BY EYEPIECE PROJECTION

Even the added focal length obtained by using your telescope mirror as a camera lens is not sufficient to give good plate scale for the most difficult group of celestial objects. These include the planets, faint nebular formations such as the Ring Nebula in Lyra, and individual craters and other formations on the moon. These can be brought to book by the addition of one more piece of equipment, an eyepiece placed between the diagonal and the camera. The eyepieces used for photographic purposes can, of course, be your telescope eyepieces. But they must be of high quality, for the film emulsion is much less tolerant of eyepiece defects than is the continuous scanning of the image by the eye.

The eyepiece increases the focal ratio—here called effective focal ratio—of the optical system according to the formula

$$efr = (\frac{L}{fe} - 1) \times \frac{fm}{a}$$

where f_m = focal length of mirror
a = clear aperture of mirror
L = distance of eyepiece from focal plane
f_e = focal length of eyepiece

M-57, the Ring Nebula in Lyra. Twenty-minute exposure on
103a-F, developed 6 minutes in D-19. Eyepiece projection, using
1-inch eyepiece on 12-inch f/4.3 reflector. (Photo by G. T. Keene)

If you use a 1-inch eyepiece placed 4 inches from the film plane in
combination with a 56-inch focal-length mirror of 8-inch diameter, the
effective focal ratio becomes

$$efr = \frac{(4-1)}{1} \times \frac{56}{8} = 21$$

which is equivalent to increasing the f-number of the telescope mirror
3 times. If the f-number increases 3 times, so will the equivalent focal
length of the telescope optical system, and also the plate scale. Put in
simplest terms, the plate scale will be increased by introducing an eyepiece into the system, and the amount of increase will be equal to the
distance of the eyepiece from the focal plane divided by the focal length
of the eyepiece, less one. Thus a ½-inch eyepiece placed 4 inches from the
focal plane will increase the plate scale 7 times, and the images on the film
will be 7 times larger.

This is of great importance in photographing the planets because it
produces a sizable image on the emulsion which needs little enlarging.
Now it is possible to get good pictures of Saturn, Jupiter, globular clus-

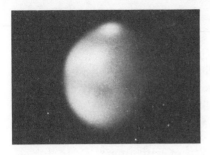

September 25, 1:41 UT. Central meridian 15°, latitude at disk center −16°. Four ½-second exposures.

September 26, 23:32 UT. Central meridian 325°, latitude of disk center −16°. The cloud extends eastward over Hellas and Ausonia. Six ½-second exposures.

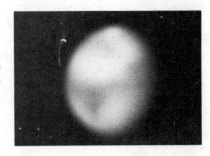

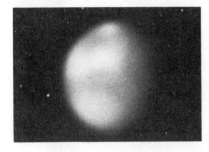

September 26, 1:20 UT. Central meridian 1°, latitude of disk center −16°. The bright cloud has a well-defined western edge. Part of Pandorae Fretum is clear. A narrow streak of cloud extends northwestward. Three 1-second exposures.

September 27, 1:08 UT. Central meridian 348°, latitude of disk center −16°. The cloud extends over Hellas and Noachis. Syrtis Major is on the left edge and Sabaeus Sinus is at the center, with a very dark spot on the Pandorae Fretum to the south. Six ½-second exposures.

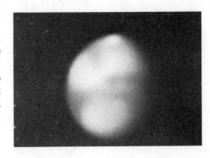

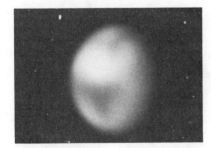

October 2, 2:37 UT. Central meridian 323°, latitude at disk center −17°. The South Polar Cap is covered, Sinus Sabaeus is gone, as is a bright cloud over Aeria to the right of Syrtis Major. A yellow K-2 filter was used. Six ½-second exposures.

Mars, during the great dust storm of 1971. Although the features are somewhat obscured, you can almost feel the rotation of the planet. The photographs were taken with a 9-inch Clark refractor, on Tri-X film at f/250 by eyepiece projection. The film was developed in DK60a, 7½ minutes at 68°, then positives were made on Contrast Process Ortho, and the composite negatives on Professional Copy. (Photos by Dennis Milon) ters, nebulae, and galaxies. Mars will still be a recalcitrant object because of its small angular diameter.

Eyepiece projection can be used with your camera if you make an adapter tube for it. The function of the tube is to keep out extraneous light when the camera shutter is opened and to provide a seat for the eyepiece. Find a length of tubing the same size as the opening in the front of the camera (where the lens elements screw in) and have it threaded to fit the opening. Obtain a piece of hard rubber about 1 inch thick—a hockey puck is excellent for this—and cut a disk from it which will just fit the adapter tube. Cut a $1\frac{1}{4}$-inch hole in the center of the disk to hold the eyepiece. Slide the disk and eyepiece into the adapter tube with the eye lens of the eyepiece directed toward the film plane until the assembly is the distance from the film plane that you desire. Usually 4 inches is a good distance for this: a $\frac{1}{2}$-inch eyepiece will increase the plate scale 7 times, and a 1-inch eyepiece will increase it 3 times. Secure the hard rubber disk in place with a set screw through the side of the adapter tube. Place the camera on the bracket rod just as you did for prime-focus photography and focus it by moving it up or down the rod. Again, if you wish more delicacy in focusing, you will have to make a coupling to attach the adapter tube of the camera to the draw tube of the telescope eyepiece holder. This makes it possible to use the telescope focusing device to focus the camera.

Exposure time for planetary and moon shots using eyepiece projection is short, ranging from $\frac{1}{10}$ to 10 seconds. Moon photography, especially when the moon is full, should be done at the shorter end of the exposure scale. Slow film of the Plus-X range can be used for eyepiece projection throughout the range of illumination of the objects photographed, whether it be the brilliant moon or the dimmer planets. If slow film is used, it should be forced in development. If you use slow film for the planets at exposures of 3 to 5 seconds, you must also use a developer which will increase the ASA exposure index to the neighborhood of 200. Color areas on the planets can be heightened in detail by the use of Anscochrome or similar film if this too is forced to speeds of 250 to 300.

For nebulae and galaxies, plate scale and exposure time matched to the size and brightness of the object must be calculated in advance. Unfortunately, these advance calculations are usually made on information gathered from past mistakes. As a starting point, a bright nebula such as the one in Orion can be photographed at an exposure of 10 minutes with Tri-X. The dimmer ones will vary from 20 minutes to 2 hours. Needless to say, the strain of guiding a telescope for periods longer than 20 minutes is a severe one, and you will attempt the dimmer nebulae and galaxies only if you become severely bitten with the photographic bug.

EYEPIECE PROJECTION PLUS CAMERA LENS

For the largest possible plate scale, you can use the eyepieces mentioned in the previous section plus the camera lens. A formula which gives the effective focal ratio for the telescope-mirror–eyepiece–camera-lens combination is

$$efr = \frac{f_c \times f_m}{a \times f_e}$$

where f_c = focal length of camera lens
  f_m = focal length of telescope mirror
  a = clear aperture of mirror
  f_e = focal length of eyepiece used

In this combination the camera lens, focused to infinity, is placed behind the eyepiece and close to it. Just how close is a matter of trial and error, but eyepiece and lens should be nearly touching.

In this type of photography the settings for exposure, the films and developers used, and general procedure follow closely those for ordinary eyepiece projection. This general statement must be qualified to the extent that any additional optics placed between camera and mirror cut down the light which reaches the film, and all the elements involved must be modified accordingly.

The increased scale makes it possible to get pictures of objects which are very difficult by other methods. By using a suitable combination of eyepiece and camera lens—say a ½-inch eyepiece and a 6-inch camera—with the 56-inch focal length of your telescope, you can obtain pictures of Mars which will exceed your hopes.

GENERAL NOTES ON PHOTOGRAPHY

We place this section at the end instead of the beginning of the chapter because much of it is based on material included within the chapter.

Plate Scale

There are two important considerations which must always be applied in choosing plate scale. The first is that reflectors, especially those of short focal length, are subject to coma at the edge of the field and thus have a smaller usable field than theory predicts. The outer edges of your negatives are not of much use, photographically, so if you choose a plate scale in which the object spreads all the way across the field, its outer limits will not be sharply defined. Consequently, choose a plate scale somewhat smaller than the formulas call for and use the sharp images in the middle area of your film.

Guiding difficulties increase with plate scale. Guiding errors which

Ingenuity in a camera mounting. This arrangement by 19-year-old
Jerry Berndt, of Yakima, Washington, makes use of a Rolleicord
camera in conjunction with the eyepiece of his telescope. (Photo
by Jerry Berndt)

would pass unnoticed with a guided camera become intolerable with eyepiece projection. For high values of plate scale you need the best finder on
your telescope that you can obtain, used at the same magnification as that
at which the photograph is taken.

Atmospheric Conditions

Celestial photography is possible only under the best possible conditions of "seeing." Atmospheric turbulence must be at a minimum. An image which jumps around in the field because of wiggling layers of atmosphere can be followed by the eye without much difficulty, but the camera does not have this facility and the result is a blurred picture. The "twinkling" stars of a cold winter's night are therefore poor subjects for the camera. Summer brings the best atmospheric conditions for photography, but along with it comes the increased sky background light because of the longer twilight.

Finally, although the stars and planets look alike to the naked eye, they don't present this uniform appearance to a photographic emulsion. They differ in color, in size, in the amount of light produced or reflected, and in their position in the sky. This latter point is important. Stars directly overhead are much easier to photograph than those near the horizon because their light passes through less dense layers of atmosphere. To cover all these points in a discussion of celestial photography would require a good-sized volume. This brief chapter is meant only to be an introduction to the subject and is presented in the hope that if you use it as a starting point, your own interest will carry you the rest of the way.

CHAPTER **18**

Other Telescopes

IF YOU ARE LIKE MANY AMATEURS, the completion of your 8-inch Newtonian reflector and the satisfaction you obtain from its performance will probably provide the impetus to start work on your next telescope. Depending on your ambition, the amount of time you have available, and the amount of equipment you have gathered at this point, you will consider one of the following as your next field of endeavor.

1. A specialized Newtonian—a long-focus instrument (Fig. 1) for planetary or lunar work, or one of short focus for observing large areas of the sky (Fig. 2).
2. A larger Newtonian. This is the telescope usually constructed as a second instrument. You are perfectly satisfied with your first one and yet—well, you would like to try something a little larger, or a little more difficult. Or to get closer to the truth, you like making telescopes and you're looking for an excuse to make another.
3. A compound telescope. You want a high-powered instrument, but the Newtonian which will give you this power is just too large for your back yard. You have read that by adding a curved secondary mirror to the ordinary paraboloidal primary, you can achieve this power and still have a compact, easily handled instrument. So you decide to build a Cassegrainian (Fig. 3), which has a convex secondary mirror, or a Gregorian (Fig. 4), which has a concave secondary.
4. A catadioptric telescope. An intriguing problem in telescope making. This type of instrument combines the reflector and the refractor.

308

It is a reflector because it employs a primary mirror, and also a refractor because of the correcting plate or lens placed at the front of the tube. If the catadioptric is to be used for photographic purposes only, you will build a Schmidt (Fig. 5) which will enable you to take pictures of wide areas of the sky with beautiful fidelity. If you want both visual and photographic use, the Maksutov (Fig. 6) will be your choice.

5. A combination telescope. The catadioptric is capable of almost endless variation of form. By adding various combinations of secondary mirrors—flat, convex, or concave—you may transform a single telescope into a multipurpose instrument.

6. A refractor. If you attempt one of these, you will have shifted completely from mirror making to lens making.

It is impossible to give detailed accounts of all these possible telescopes —the Maksutov alone would require a fat volume to be described adequately—but we can go into some of their characteristics and discuss briefly how they are made.

LONG-FOCUS NEWTONIAN REFLECTOR

Here is the most beautiful definition possible in a reflector, and although the mirrors are difficult to figure properly, the results are so rewarding that the effort is worth while. Paraboloidal mirrors of f/9 and f/10 vary little from spherical mirrors of the same f-ratio, yet this small difference is all-important in performance.

The source of the difficulty lies in the intensity of the shadows in testing for the paraboloid. In the f/7 mirror you can differentiate between shadow patterns easily, for their boundaries are reasonably distinct. But in the higher ratios the shadows are faint and extremely delicate. Just seeing them is difficult enough; measuring them is an art rather than a science. The mirror maker must be plentifully endowed with patience; he must approach the final stages of polishing with great care; he must be prepared to retrace his steps in spite of this approach. If he parabolizes the mirror until he can see the shadows clearly, he must return to the sphere again, for at this point he has grossly overcorrected the figure.

The alternative is to increase the focal length still more, say to f/12. Now the mirror may be left spherical because the difference between sphere and paraboloid is negligible at this ratio. But at f/12, a 6-inch mirror will have a focal length of 6 feet; an 8-inch mirror must have a tube at least 8 feet long. A tube this long requires a mounting of *great* rigidity. Your choice now is to build a mounting as solid as the rock of Gibraltar or to spend the time saved by a careful figuring of a slightly shorter focal length mirror. Most workers prefer the latter.

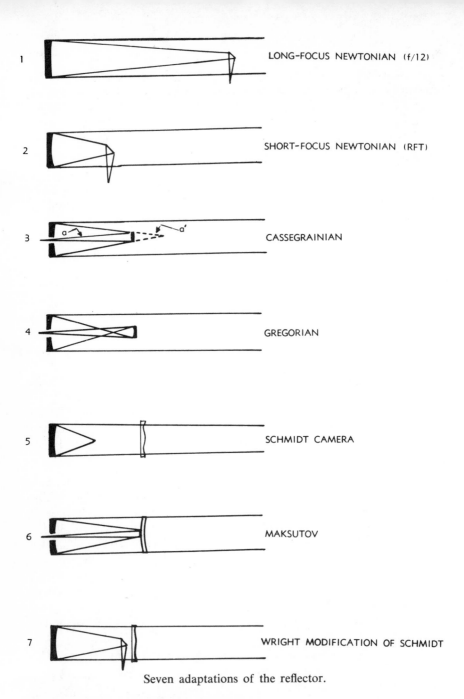

1	LONG-FOCUS NEWTONIAN (f/12)
2	SHORT-FOCUS NEWTONIAN (RFT)
3	CASSEGRAINIAN
4	GREGORIAN
5	SCHMIDT CAMERA
6	MAKSUTOV
7	WRIGHT MODIFICATION OF SCHMIDT

Seven adaptations of the reflector.

SHORT-FOCUS NEWTONIAN REFLECTOR

In the short-focus telescope (f/5 and below) we are sacrificing definition for width of field, because coma is an increasing defect as the focal length becomes smaller. In exchange, we obtain breath-taking views of open star clusters, wide-spread galaxies and extended nebulae. An examination of the Milky Way through a telescope of this type is never to be forgotten, for it is only by this means that one can grasp the fact that the stars are really numberless. A name popularly applied to the short-focus telescope is RFT, or rich-field telescope.

In reflectors, the best focal ratio for a 6-inch RFT is 4, for a variety of reasons.[1] Working from the "seeing" end of the telescope to the light-gathering end, we start with the eye itself. The pupil of the eye, at night, averages 7 mm in width. The telescope must produce a cone of light, or exit pupil, just this size. If the cone is too large or too small (see page 151) the eye cannot be used at its full efficiency. The formula which determines the size of the exit pupil is:

$$d = \frac{D \times f_e}{f_m} \text{ or } \frac{D}{d} = \frac{f_m}{f_e}$$

where D = diameter of mirror
d = diameter of exit pupil
f_m = focal length of mirror
f_e = focal length of eyepiece

We choose a 6-inch mirror because this size is the easiest to grind, polish, and figure for the deep curve that we must produce (the sagitta for a 6-inch f/4 is .0937 inch). Assuming a 1¼-inch focal length eyepiece (most easily obtained in wide-field eyepieces), we obtain

$$\frac{7}{25.4} = \frac{6 \times 1\frac{1}{4}}{f_m} \text{ , or } f_m = 24.6 \text{ inches}$$

But the focal length of the mirror divided by the aperture equals the focal ratio, or 24.6/6 = 4.1 (see page 6). The magnification of the system equals 24.6/1¼, or about 20.

Now suppose the 1¼-inch eyepiece is a Kellner type with an apparent field of 50°. Since the real field of a telescope equals the apparent field divided by the magnification, this telescope will cover 2.5° of the sky (see page 150).

The combination described above will pick up stars as faint as 12th magnitude on a night of good seeing. Because of its low magnification and light weight it can be carried under the arm or, if mounted, it does not need as solid support as other telescopes. And because of its wide field, it needs no finder.

It is a common misconception that because the RFT is designed for low-power wide-field observation, it needs less attention in figuring and collimation than other telescopes. Nothing could be further from the truth.

It is true that even a perfect paraboloid of the short focal length involved here cannot bring the rays of an off-axis star to a sharp focus, and

[1] See Albert G. Ingalls (editor), *Amateur Telescope Making,* Book II, Scientific American Publishing Company, New York.

we must expect that the star images on the outer edges of the field will not be as sharp as those near the center. But this is no excuse for sloppy work in parabolizing. In fact, it should be the reason for even more rigorous efforts to obtain a good figure. The same effort must be made to collimate the optical train of the telescope. Mirrors of low focal ratio must be aligned carefully; a very small error produces a large amount of distortion in the images. Finally, the diagonal must be of very good quality because the reflected rays from the mirror strike it at a sharper angle than in a telescope of longer focus.

Parabolizing a short-focus mirror is usually a long process. The deep curve in the mirror is hard to make deeper to produce the paraboloid, although the process may be hastened by using a soft lap. This is really an advantage because you approach the paraboloid slowly and thus have plenty of time and opportunity to adapt your strokes to the conditions that you find on the mirror's surface. Don't bother to make the mirror spherical before parabolizing; start your long strokes as soon as you have a good polish and work toward the paraboloid almost from the beginning.

LARGER MIRRORS

The difficulty in making mirrors increases roughly as the cube of the diameter—and some amateurs who have started with a 6-inch mirror and then jumped directly to one of 12 inches would say the ratio should be nearer the 4th power than the 3rd. Nor is this the end. Everything else in the large-mirror field increases out of proportion to the diameter. The mounting begins to assume massive proportions, and portability for any instrument over 12 inches becomes a practical impossibility. The physical labor of grinding and polishing a large mirror is prodigious.

Most workers, when they reach the stage where a "big" mirror *has* to be tackled, build a grinding and polishing machine. Such a machine duplicates as nearly as possible the motions of the mirror maker—pushing the mirror back and forth across the tool, rotating it on its axis, and rotating the tool. It is fun to make and even more fun to use. Sitting back while listening to the groans and squeaks of a machine as it grinds the mirror is a unique and never-to-be-forgotten experience for the amateur telescope maker. The fact that he substitutes the anguish of wondering whether the machine is doing the job right for the physical labor of doing it himself is beside the point.

Polishing, too, brings its problems. A 12-inch mirror has over twice the area of an 8-inch, but there seem to be more than twice as many zones which can act up. It is very difficult to keep the lap in good contact over such a wide area, and irregularities in the mirror's surface are common. Here is where the sub-diameter lap really comes into its own. The job of figuring the mirror to the paraboloid also has its complications. Many zones must be tested to make sure that the curve is regular from center to edge. The Foucault test can still be used effectively, but many mirror makers use it chiefly for the general curve, supplementing it with more specific testing devices, such as the caustic and Hartmann testers.[2]

These difficulties deter very few amateurs, however, and a vast number of mirrors of 12-inch size and over have been successfully completed. Still, the jump from 8-inch to 12-inch is a large one, and it is best to work on at least one other optical surface—say a 6-inch RFT or another 8-inch of different focal length from your original to gain some more experience. When you have reached the point where a 12-inch or larger is indicated, consider all aspects of the problem before you begin. For example, an f/10 12-inch mirror may appear very desirable on paper because the long focal length seems ideal for planetary observation, but when you start to build a mounting for a 10-foot tube you may decide otherwise.

THE CASSEGRAINIAN

People who have encountered the problem mentioned in the paragraph above, or who find the eyepiece position of a Newtonian awkward and tiresome, are inclined to look at the Cassegrainian telescope as the answer to their troubles. To increase the magnification in a Newtonian, you must increase the focal length of the mirror, decrease that of the eyepiece, or use a Barlow lens in the system. There are disadvantages in each of the three possibilities. But the Cassegrainian not only yields high magnification with a short tube; it also has the eyepiece placed where you may reach it comfortably.

The principle of the Cassegrain is the same as that of the Barlow lens. The cone of light reflected from the primary mirror is cut off at some point before it reaches the focal plane and is replaced by another, but longer, cone. The length of the new cone divided by the part cut from the original one equals the amplification, or increase in magnification, of the system. In the Barlow lens, the cone is lengthened by means of a negative lens placed in the optical train; in the Cassegrain another mirror is introduced. Just how this works is best shown by looking at the diagram (Fig. 3). Let us assume cone a to be 8 units long and cone a' 2 units. The amplification factor is then 4. If the original focal length of the primary mirror is 32 inches, the equivalent focal length of the whole system now becomes 128 inches.

All that we need do to increase the magnification of a mirror, then, is to drill a hole through its center and place a convex mirror somewhere

[2] "The Caustic Test" by Irvin H. Schroader in Albert G. Ingalls (editor), *Amateur Telescope Making*, Book III, Scientific American Publishing Company, New York, pages 429-456; and "The Hartmann Test" by William A. Calder in *Amateur Telescope Making*, Book II, pages 109-112.

in the optical train. This sounds very simple; and, if it really were this easy, most of the telescopes today would be Cassegrainians. The stumbling block is how to make and test the convex secondary mirror.

The primary mirror is a paraboloid, as in the Newtonian, but the secondary must be hyperboloidal. Up to a few years ago, the only way known to test such a mirror was to construct an optical flat at least the size of the primary mirror. But making an optical flat of this size is at least as difficult as making the primary mirror itself. In other words, to make a Cassegrain it was necessary to grind and polish three optical surfaces, one concave, one convex, and one flat.

Because of the difficulty in testing the hyperboloidal mirror, amateurs looked around for other systems which would perform as well as the Cassegrainian but whose optics were easier to produce. Perhaps the most successful of the resulting telescopes was the Dall-Kirkham, in which the problem was somewhat simplified by substituting a spherical secondary for the hyperboloid and undercorrecting the paraboloid on the primary mirror. A spherical curve is the easiest to grind and figure on a convex surface. The amount of undercorrection for the primary mirror can be determined by testing it in conjunction with the convex secondary. It varies between 60 and 75 per cent. Neither of the mirrors requires an optical flat for a test, consequently the Dall-Kirkham is much simpler to construct and test than the Cassegrainian.

Other compound telescopes include the Gregorian, in which an elliptical concave secondary is placed in the optical train at a point beyond the focal point of the paraboloidal primary (Fig. 4). The Schwarzchild uses the same system of concave mirrors except that the secondary is moved inside the prime focus. Finally, the Ritchey-Chretien is similar to the Cassegrainian in that it makes use of two aspheric surfaces, one convex and one concave, the secondary a deformed hyperboloid and the primary a deformed paraboloid. Compared to this, figuring a Cassegrainian is easy.

None of these systems, with the exception of the Dall-Kirkham, is of much interest to the amateur unless he is deliberately looking for new and difficult worlds to conquer. Nor does he need to avoid the Cassegrainian hyperboloid any longer, for recent developments in testing techniques have solved this problem, too. A simple test was devised and published in 1939 by Enrique Gaviola.[3] For some reason this simplified method was overlooked by amateurs and it is only very recently that it has come into its own. The only apparatus required is a reading glass and a flat the size of a Newtonian diagonal. Space does not permit a full discussion of the Gaviola test here; but, for those who are interested, a brief description is included in appendix VII, pages 336-337.

[3] *Journal of the Optical Society of America*, November 1939, volume 29, pages 480-483.

All the compound telescopes require a perforated primary made by the biscuit cutter described earlier. The mirror must be drilled from back to front during fine grinding to within $\frac{1}{16}$ inch of the surface. If this is done *after* the mirror is completed it relieves strains in the glass, and raised zones around the hole are produced. Beeswax is used to fill the cut, after which fine grinding, polishing, and figuring are completed. Finally, the hole is cut through to the surface.

CATADIOPTRIC SYSTEMS

By definition, a catadioptric system is one which combines mirrors and lenses in the optical train at some point before the light reaches the eyepiece. Strictly speaking, an ordinary Newtonian becomes a catadioptric telescope if you add a Barlow lens to it, but the term is usually reserved for those systems which employ a correcting plate at the front of the telescope. There are many variations of the combination of correcting plate and primary mirror, only one or two of which are of interest to the amateur.

The Schmidt Camera

The primary mirror in the Schmidt is spherical and, although its deep curve is difficult to grind, offers no more complications than any other short-focus mirror. The correcting plate is quite another matter because it must be perfectly flat on the mirror side and be ground to a complicated curve on the other. The image is formed between mirror and correcting plate and has the great disadvantage of lying on a curve, so that the film used must also be curved. Once these difficulties are overcome, however, the Schmidt is a wonderful photographic instrument for wide-field work, and the many amateurs who have made Schmidts consider the results obtained as more than worth the trouble.

The Schmidt itself cannot be used visually, although an adaptation of the Schmidt system by F. B. Wright makes it possible to employ a Newtonian diagonal for visual work. In the Wright system the primary mirror is ground to an oblate spheroid, the diagonal is placed immediately behind the correcting plate, and the result is a flat field which may be used both visually and photographically.

The Maksutov Telescope

Because of its many qualities and relatively few drawbacks, the Maksutov is probably the most popular catadioptric telescope with amateurs. The correcting plate is spherical on both sides, concave toward the front of the telescope and convex on the mirror side. The primary mirror is also spherical, although much flatter than the other two surfaces. One form of Maksutov, for example, has a primary whose focal length is approximately 5 times either of the two curves on the correcting plate.

The Maksutov spherical surfaces may be used in many ways. The telescope may become a form of Newtonian by the addition of a flat diagonal, a Cassegrainian if the secondary mirror is a convex hyperboloid, or a Gregorian if concave ellipsoid is used. The most popular Maksutov is the Cassegrainian adaptation. Compactness is increased when the secondary mirror is an aluminized spot on the back of the correcting plate, as shown in fig. 6.

Building a Maksutov is not a project to be approached lightly, even though the problem of grinding three spherical surfaces may not seem insurmountable. But these must be *matching* surfaces, each curve having specifications which must be observed exactly. Furthermore, the thickness of the correcting plate and the index of refraction of the glass employed, as well as its optical qualities, have to be taken into consideration. Finally the correcting plates, although now more available than formerly, have been expensive and difficult to obtain.

None of these difficulties is likely to deter amateurs. The Maksutov Club, founded by Allan Mackintosh for the purpose of disseminating information to its members and making it possible to buy corrector blanks in quantity, had two hundred members within months of its founding. The reason for the enthusiasm is that the Maksutov is perhaps the ultimate in telescopes. It outclasses reflectors and refractors in almost every respect. With a tube only 18 inches long, it can have an equivalent focal length of 100 inches. Because of the short tube, an equatorial mounting is easy to make and use. The optical characteristics of the system are of a high order. It has freedom from coma and astigmatism and has a beautiful flat field. Because it has a closed tube, thermal effects are almost negligible. When the Cassegrainian secondary is a part of the correcting plate, there are no diffraction effects from spider arms. Finally, because the telescope has only two optical parts, each of which is firmly seated in its own cell, it is exceptionally rugged and will stand much rough treatment.

REFRACTORS

Making the optics for a good refractor is perhaps the most difficult task the amateur can approach. He must be prepared to grind and polish *four* optical surfaces, each with its own figure and, as in the catadioptric systems, these must be *matching* surfaces. He must acquaint himself thoroughly with the optical qualities of various glasses, for he is now using the interior of the glass as well as its exterior surface. He must learn how various glasses complement each other for color correction. These, combined with new grinding and polishing techniques, new testing devices, and other skills which are required in lens manufacture, make the refractor a formidable problem for the amateur.

But for the worker who likes to fabricate optical parts into telescope units, the refractor is a stimulating and challenging opportunity. Objectives of good resolution and small spherical and chromatic aberration can be obtained from several of the war-surplus suppliers or direct from manufacturers. Good achromatic object lenses are expensive: a 4-inch f/15 costs from $75 to $125, and a 6-inch up to $500. On the other hand, a finished 6-inch refractor, complete with all accessories, may cost seven or eight times this much. The job at hand is to mount and collimate the optical elements in a tube, then to produce a mounting for the tube. It must not be thought that this is a simple operation, but people who are skilled at machine-tool operations will find the job fascinating.

A SUMMARY

Any of the telescopes mentioned above can be made by the determined amateur. They are graded in roughly increasing difficulty. If you have found making your first telescope fun, making some of these other types provides a hobby of almost endless possibilities. Each represents a notable achievement in its own right. The literature on these telescopes is voluminous; a bibliography is included in the appendix for your convenience. Good luck!

Glossary

AAVSO   The American Association of Variable Star Observers.

Accelerator   A compound that speeds up the setting of a polyester resin, usually cobalt naphthenate.

Achromatic lens   A two-element lens which theoretically produces a color-free image (not completely possible in practice).

Airy Disk   The disklike image of a point source of light as seen in an optical system.

Aloxite   See Alundum.

ALPO   The Association of Lunar and Planetary Observers.

Altazimuth mounting   A mounting consisting of two axes at right angles to each other. One axis is vertical, the other horizontal.

Altitude   The angular distance of a celestial body from the horizon, measured along a line which passes through the zenith.

Aluminizing   The process of coating a mirror with a thin layer of vaporized aluminum.

Alundum   Trade name for grinding compound made from fused aluminum oxide.

Angular diameter   The apparent diameter of an object measured in degrees.

Angular motion   The motion of a celestial body measured in degrees.

Apparent motion   The motion which the stars seem to have because of the rotation of the earth.

Aries   A constellation (popular name, The Ram) near which the vernal equinox is located.

Astigmatism   Any distortion of an image resulting from the failure of rays from different parts of the same zone to focus in the same plane.

Azimuth   Lateral direction, measured in degrees from north clockwise around the horizon. Thus, due west has an azimuth of 270°. More simply, it is the bearing or direction of a body.

Back focal length   The distance from the last surface of an eyepiece to the focus of the emergent rays.

Barlow lens   An intermediate lens placed in front of the eyepiece of a telescope to increase magnification.

Barr scale   A scale used to measure the longitudinal motion of the knife edge in a Foucault testing device.

Beral coating   An alloy of beryllium and aluminum used to provide a reflective coating on mirrors.

"Biscuit" cutter   A large-sized drill made of a piece of tubing mounted on a shaft; used to drill holes in mirrors.

Carborundum   A grinding compound made from silicon carbide, one of the hardest substances known.

Cassegrainian   A compound telescope that has a paraboloidal primary mirror and a hyperboloidal secondary.

Catadioptric telescope   A compound telescope combining optics of refractor and reflector. It employs a correcting lens as well as a mirror.

Catalyst   Any substance that speeds up a chemical reaction. The one used with polyester resins is usually methyl ketone peroxide.

Cepheids   Variable stars useful in measuring astronomical distances. So named because the first one found was located in the constellation Cepheus.

Cerium oxide   A polishing agent made from the rare earth element cerium.

Chromatic aberration   The breaking up of "white" light into its colored constituents. It is characteristic of lenses; not present in reflecting telescopes.

Chromatic difference in magnification   The tendency for a lens to magnify some colors more than others, especially at the edges of the field.

Cold camera   A camera containing an extra chamber for dry ice to increase the sensitivity of film.

Collimate   To line up the optical elements in a telescope.

Coma   The flaring of a star image into an umbrellalike distortion.

Conic section   The curve produced when a cone is cut at an angle.

Coordinates   Two or more values locating a point in space.

Correction   The amount of deviation of a mirror surface from a perfect paraboloid. Overcorrection implies too deep a curve; undercorrection too shallow.

Correcting lens   The lens placed at the front of a catadioptric telescope; used to correct spherical aberration of the primary mirror.

Corundum   Aluminum oxide found in nature.

Couder screen   A mirror mask provided with holes cut at certain intervals for testing purposes.

Crest   The darkest part of a shadow pattern on a mirror.

Curvature of field   The apparent bending of the field, requiring a difference in focusing for the edge areas.

Dall-Kirkham telescope   A variation of the Cassegrainian telescope. It makes use of a spherical secondary mirror and an oblate spheroid for a primary.

Dawes' Limit   The smallest angular separation of two stars, in which each is still observable with a telescope of given aperture.

Declination   The angular distance of a heavenly body above or below the celestial equator.

Declination axis   The axis about which a telescope is turned to make adjustments in declination.

Declination circle   A setting circle divided into units of 0° to 90° to measure declination.

Definition   Faithful reproduction of the characteristics of an object over the whole field. Definition depends on the quality of the mirror.

Deflocculant   A substance which, when added to a suspension, will prevent clumping of individual particles into larger masses.

Detergent   A "soapless" soap; one which does not react with minerals in the water and which also breaks up surface tension to make suspended particles "wetter." Examples: Dreft, Glim, etc.

Diagonal mirror   A small flat mirror used to reflect the main cone of light into the eyepiece of a Newtonian telescope.

Diffraction   The spreading of a beam of light into the shadow, causing a light and dark pattern because of the interference of light waves.

Diffraction rings   The rings that surround the image of a star or other point source of light.

Distortion   The curving of what should be straight lines at the edge of the field.

Dobson telescope   An altazimuth telescope noted for its simple construction, short focal length, and large aperture.

"Dog-biscuit" surface   The lumpiness that appears on the surface of a mirror under the Foucault test.

"Doughnut" shadows   The apparent ringlike shadows characteristic of the oblate spheroid, paraboloid, or hyperboloid which appear under the Foucault test.

Double star   A star that splits into two components under optical aids.

Draw tube   A freely moving tube that holds the eyepiece in a telescope.

Ecliptic   The path of the sun during its apparent annual motion around the earth.

Effective focal ratio   The ratio between the effective focal length of a compound system and the aperture of the mirror.

Emery   A grinding compound that is a mixture of the oxides of aluminum, silicon, and iron.

Ellipse   An oval figure in which the sum of the distances from two points (called the foci) to any point on the curve is always equal to a fixed quantity.

Ellipsoid   The three-dimensional figure produced by revolving an ellipse around its axis.

Ephemeris   A book of tables of daily positions of heavenly bodies.

Equatorial mounting   A telescope mounting in which one axis (polar) is parallel to the axis of the earth, with a second axis (declination) perpendicular to the first.

Equinoctial   The celestial equator.

Equinoxes   The two points at which the sun crosses the equinoctial.

Equivalent focal length   The useful focal length of a compound optical system.

Everest test method   A test method for checking the surface of a mirror in which the extent and motion of shadow patterns is examined.

Eye lens   In an eyepiece, the lens closest to the eye.

Eyepiece projection   The system in celestial photography in which an eyepiece is used to enlarge the image falling on the film.

Eyepiece test   A test for turned-down edge on a mirror, in which an eyepiece replaces the knife edge of the Foucault test.

Eye relief   The distance the eye must be placed from the eye lens to obtain sharpest vision.

Exit pupil   The size of the emergent cone of light from the eyepiece of a telescope at the point of sharpest vision.

f-number   The numerical ratio between the focal length of a mirror or lens and its diameter; a measure of the potential brightness of the image.

Field   The area visible through a telescope.

Field lens   The lens furthest from the eye in an eyepiece.

Field of view   The angular diameter of the field covered by a telescope or other optical device.

Figure of revolution   Three-dimensional figure produced by the revolution of a curve around its axis.

First-surface mirror   A mirror in which the reflective coating is placed on the front, rather than the back, surface.

Focal length   The distance at which rays reflected from the surface of a mirror intersect each other.

Focal plane   The area in which the reflected rays from a mirror intersect to form an image.

Focal ratio   Synonymous with f-number.

Fork mounting   A telescope mounting in which the tube is mounted between two prongs of a fork, the shaft of which is supported on bearings.

Formica   Trade name of a very hard-surfaced material used where a smooth friction-free surface is needed. Often a component of Dobson telescopes and Poncet mountings.

Forming gas   Made up of 8 per cent hydrogen and 92 per cent nitrogen. Used to increase film sensitivity.

Foucault test   A method of testing mirror surfaces by cutting reflected rays with a knife edge and observing the shadows produced.

Galaxy   A great system of stars, such as our own Milky Way system or the tremendous group visible in the constellation Andromeda.

Galilean moons   Four moons of Jupiter first observed by Galileo.

Galilean telescope   A refracting telescope that uses a concave lens for an eyepiece. It produces an erect image.

Garnet   Naturally occurring mineral used for fine grinding.

Gatorfoam   A light, relatively strong, spongy substance sometimes used in the construction of telescope tubing.

Gaviola test   A simplified test for Cassegrainian secondary mirrors.

German mounting   An equatorial mounting in which the two axes meet each other in a tee.

Ghost image   Nebulous areas of light that can be seen in certain types of eyepieces.

Globular cluster   A group of stars so spaced that they give the impression of a globe.

Greenwich Civil Time   A time system based upon twenty-four time zones originating at Greenwich, England; often called Greenwich Mean Time or Universal Time.

Greenwich meridian   The meridian of longitude that passes through Greenwich, England.

Gregorian telescope   A compound telescope that makes use of a concave secondary mirror.

Grit   Term used by telescope makers to refer to abrasives.

Guided camera   A camera attached to an equatorial mounting for the purpose of photographing celestial objects.

Hartmann test   A test used primarily for large mirrors, where the Foucault test is less effective. The mirror is tested by means of a screen in which holes are punched along various diameters. The holes are then photographed inside and outside focus and the results compared to find the actual focal point of any given zone.

Herschel wedge   A wedge-shaped prism used for observing the sun. It permits only a small number of solar rays to reach the eyepiece.

Hour angle   The angular position of a celestial body east or west of the meridian.

Huygenian eyepiece   A negative eyepiece in which the focal plane lies between the eye and the field lens.

Hyperbola   An open curve obtained by cutting a cone with a plane that makes an angle with the base greater than that made by the cone's side.

Hyperboloid   The three-dimensional figure obtained by revolving a hyperbola around its axis.

Hypersensitizing   The process of increasing the sensitivity of film through low temperatures or the use of forming gas.

Interference fringe   The dark lines obtained by pressing two pieces of flat glass together, caused by alternate interference and reinforcement of light waves.

Kellner eyepiece   A positive eyepiece in which two plano-convex lenses are used for field and eye lenses. The convex surfaces are oriented toward each other.

Knife edge   Term used by mirror makers for any sharp edge used to cut a cone of light. It is usually a razor blade.

Konig eyepiece   A four-element eyepiece noted for its clear, sharp field. Contains a cemented achromat and two single-element lenses.

Lap Any surface upon which a mirror may be polished. It must be yielding enough to embed the polishing agent in its surface. Common materials used are pitch, beeswax, plastic, silk, paper.

Latitude (celestial) The angular distance from the ecliptic, as opposed to declination, which is the angular distance from the equinoctial.

Lemon-peel surface The surface of a mirror on which the imperfections have a resemblance to the skin of a lemon.

Lens A curved piece of glass used to bend light rays toward or away from one another.

Local meridian (celestial) The meridian that happens to be overhead at any instant of time.

Local meridian (terrestrial) The meridian passing through the geographical position of the observer.

Longitudinal displacement The displacement of the intersection of a pair of light rays toward or away from the mirror.

Mackintosh tester A precise adaptation of the Foucault test device.

Magnesium fluoride A material which, when coated on a lens, increases the amount of light transmitted.

Magnification Increase in apparent size of an image.

Maksutov A compound telescope that uses three optical elements: a correcting lens, a spherical primary mirror, and a hyperboloidal secondary.

Mesh size The number of openings per square inch in the screening used to grade abrasives.

Messier objects The "false comets" Charles Messier listed to avoid confusion with the real thing.

Micron A unit used in small measurements; equals .00004 inch.

Monocentric eyepiece A single-element eyepiece sometimes consisting of a thick lens (the Tolles). The Hastings type has another lens cemented to the thick one.

NASA National Aeronautics and Space Administration.

Nebula The term applied to gas clouds among the stars, and also to some of the star systems beyond our galaxy.

Negative eyepiece An eyepiece that forms an image between the lens elements.

Newton's Rings Circular interference fringes.

Newtonian reflector Reflecting telescope using a paraboloidal primary and a flat diagonal.

Objective The principal (front) lens of a refracting telescope.

Oblate spheroid The three-dimensional figure produced by the revolution of an ellipse around its long axis.

Occultation The eclipsing of one heavenly object by another, as when the moon passes in front of a star.

Optical axis The perpendicular to a tangent drawn at the center of a lens or mirror.

Optical train The path followed by an axial ray as it passes through an optical system.

Orthoscopic eyepiece A compound eyepiece in which the field lens is a cemented achromat. It is notable for its flat field.

Overcoating Transparent coating, usually of quartz or other siliceous material, to protect the aluminum coating of a mirror.

Overhang The projection of a mirror over the tool at the end of a stroke; also applied to lateral projection of a mirror during a stroke.

Panderae Fretum A dark, shadowy area near the south pole of the planet Mars.

Parabola The curved figure produced by cutting a cone through its base along a plane parallel to the slope of the cone.

Paraboloid The three-dimensional figure produced by the revolution of a parabola around its axis.

Paraboloidal shadows Doughnut-shaped shadows appearing on a paraboloidal mirror when the knife edge is at its 50 per cent position in the Foucault test.

Parting tool Chisel-like instrument used for breaking glass along a scored line on its surface.

Per cent position One of the positions of the knife edge in the Foucault test.

Pitch Resin produced from coniferous trees or from by-products of coal tar.

Pits Tiny holes in the surface of the mirror resulting from large particles of abrasive.

Plano-convex lens A lens whose opposite sides are respectively flat and convex.

Plossl eyepiece An eyepiece made up of two identical lenses set very close together. Noted for wide field, excellent definition, and image contrast.

Polar axis The axis in an equatorial mounting parallel to the earth's axis.

Polishing Smoothing the surface of a mirror or lens, as opposed to abrading it, as in the grinding process.

Polyester resin   Synthetic resin commonly used as a matrix for embedding the fibrous material in fiberglass construction.

Poncet mounting   A platform-type mounting invented by Adrien Poncet. It is a variation of the usual equatorial mounting.

Positive eyepiece   An eyepiece that produces an image in back of the eye lens.

Primary mirror   The main mirror in a reflecting telescope.

Prime focus   The point at which the rays from the primary mirror of a reflecting telescope come to a focus.

Prime focus photography   Celestial photography in which the film is placed at the prime focus of the telescope.

Prism   Various shaped pieces of glass used to change the direction of the path of light in an optical system.

PVC   Polyvinyl chloride. Tough, hard plastic compound used for drainage pipes, sections of which make good bearings in Dobson telescopes.

Pyrex   Glass of low coefficient of expansion produced by adding boron compounds to the constituents of ordinary glass.

Rack and pinion focuser   Employs a movable toothed bar that bears against a geared wheel.

Ramsden disk   The cross section of the exit pupil at the point of sharpest image.

Ramsden eyepiece   An adaptation of the Kellner eyepiece. The eye lens is an achromatic lens instead of a single plano-convex unit.

Rayleigh standard   The requirement that all reflected rays from a mirror have no deviation greater than ¼ wave length of light.

Reciprocity failure   The loss of a film's sensitivity after a short period of time.

Reference flat   Glass of known flatness against which other flat surfaces may be tested.

Real field   The area covered by an optical system.

Refractor   A telescope employing an objective lens to bring light rays to a focus.

Resolving power   The power of a telescope in separating two close-together objects.

Reticle   Ruled piece of glass placed at the focal plane of a telescope. The rulings are superimposed on the image.

RFT   Richest field telescope.

Right ascension   The angular distance of a celestial body measured to the east of the vernal equinox.

Right ascension circle   A circle placed on the polar axis of a telescope to aid in lining up the instrument with the right ascension of a celestial object.

Ronchi test   A derivative of the Foucault test, in which the knife edge is replaced by a screen or grating.

Rouge   Form of iron oxide used in polishing optical surfaces.

Sagitta   The depth of curve in a mirror.

Schmidt camera   A photographic instrument using a correcting place and concave primary mirror. It produces a curved focal plane.

Secondary mirror   The mirror, curved or plane, that reflects light into the eyepiece.

"Seeing"   The measure of the quality of a telescope image with respect to atmospheric turbulence.

Setting circle   The circular scale used in a telescope to set it to the declination and right ascension of a celestial object.

Sidereal day   The time interval between two successive passages of a star across the local meridian. It equals approximately 23 h 56 m of solar time.

Sidereal rate   The rate at which the driving apparatus of a telescope must be set to keep a star centered in the eyepiece.

Sky transparency   The clarity of the atmosphere in terms of the amount of dust particles and water vapor present.

Sleeks   Minute, hairlike scratches on a mirror arising from the use of a very hard pitch lap.

Slump   The tendency of the edge facets of a lap to fall away from good contact with a mirror.

Solar day   The interval between two successive passages of the sun across the local meridian.

Solid eyepiece   An eyepiece in which the eye and field lenses are fused into a solid element.

Sonotube   Trade name of large tubes used in the construction industry as forms for concrete piers or footings. Made of paper impregnated with plastic. They make excellent telescope tubes.

Speculum metal   An alloy of copper and tin used by early astronomers for mirrors.

Sphere or spheroid   A three-dimensional figure produced by the revolution of a circle around its axis. Actually, "spheroid" is incorrect as a synonym for sphere, since Webster defines it as "a figure almost a sphere," but the two terms are interchangeable in mirror-making.

Spherical aberration   The deviation of reflected rays from a spherical surface from a common focal point.

Spider   The supporting structure for the diagonal or secondary mirror in a reflecting telescope.

Suspension   A mixture of water and an insoluble substance, such as an abrasive, in which the undissolved particles float in the water.

Synchronous motor   An electric motor whose rate is determined by the regularity of the cycles of alternating current that supplies its power.

Syrtis Major   A wedge-shaped dark area near the Martian equator.

Template   A curve of known radius used to measure the depth of curve of a mirror.

Thermal expansion   The expansion of a mirror with an increase in temperature.

Tool   The glass disk upon which the mirror is ground.

Toroid   A surface generated by the rotation of a plane closed curve about an axis lying in the plane.

Turned edge   The flattening of the edge of a mirror from the paraboloidal curve.

Universal time   A commonly used designation for Greenwich Civil Time.

Variable stars   Stars whose brightness varies in magnitude over periods of time.

Vernal equinox   The point in the equinoctial through which the sun passes in the spring. It is also called the First Point in Aries.

Wave length   The distance between two consecutive crests in transverse waves of light. Unless a particular color is specified, the length is taken to be that of a wave of yellow-green light, or .000022 inch.

Wet   The period in which a charge of abrasive is broken down and must be replaced.

Yoke mounting   A form of equatorial mounting in which the polar axis consists of a yoke in which the telescope tube is mounted.

Zone   A definite area on the mirror, measured from the center.

Zoom eyepiece   An eyepiece in which magnification varies with the focus.

APPENDIX I

Conversion of Mean Solar Time Interval to Sidereal Time Interval

	Correction			Correction
Hours	Min	Sec	Min	Sec
1		10	1-3	0
2		20	4-9	1
3		30	10-15	2
4		39	16-21	3
5		49	22-27	4
6		59	28-33	5
7	1	09	34-39	6
8	1	19	40-45	7
9	1	29	46-51	8
10	1	39	52-57	9
11	1	48	58-60	10
12	1	58		
13	2	08		
14	2	18		
15	2	28		
16	2	38		
17	2	48		
18	2	57		
19	3	07		
20	3	17		
21	3	27		
22	3	37		
23	3	47		

Example: What sidereal time interval corresponds to a solar time interval of 13 hours 22 min?

From table:

$$
\begin{array}{rl}
13 \text{ hours} & = \ 2 \text{ min } 8 \text{ sec} \\
22 \text{ min} & = \ \underline{\ \ \ \ \ \ \ \ 4 \text{ sec}} \\
& \ \ \ \ 2 \text{ min } 12 \text{ sec}
\end{array}
$$

or, 13 hours, 2 min, 12 sec

APPENDIX II

Sidereal Time at 0 Hours G.C.T. (Universal Time)
for the Meridian of Greenwich, 1960

Date	Jan.	Feb.	Mar.	Apr.	May	June	July	Aug.	Sept.	Oct.	Nov.	Dec.
	h m	h m	h m	h m	h m	h m	h m	h m	h m	h m	h m	h m
1	6 39	8 41	10 35	12 37	14 36	16 38	18 36	20 38	22 40	0 39	2 41	4 39
2	6 43	8 45	10 39	12 41	14 40	16 42	18 40	20 42	22 45	0 43	2 45	4 43
3	6 47	8 48	10 43	12 44	14 44	16 46	18 45	20 46	22 49	0 47	2 49	4 47
4	6 51	8 53	10 47	12 49	14 48	16 50	18 48	20 52	22 53	0 51	2 53	4 51
5	6 54	8 57	10 51	12 53	14 52	16 54	18 52	20 54	22 56	0 55	2 57	4 55
6	6 58	9 01	10 55	12 57	14 55	16 58	18 56	20 58	23 00	0 59	3 01	4 59
7	7 02	9 05	10 59	13 01	14 59	17 02	19 00	21 02	23 04	1 03	3 05	5 03
8	7 06	9 09	11 03	13 05	15 03	17 06	19 04	21 06	23 08	1 07	3 09	5 07
9	7 10	9 12	11 07	13 09	15 07	17 09	19 08	21 10	23 12	1 10	3 13	5 11
10	7 14	9 16	11 11	13 13	15 11	17 13	19 12	21 14	23 16	1 14	3 17	5 15
11	7 18	9 20	11 15	13 17	15 15	17 17	19 16	21 18	23 20	1 18	3 21	5 19
12	7 22	9 24	11 19	13 21	15 19	17 21	19 20	21 22	23 24	1 22	3 25	5 23
13	7 26	9 28	11 23	13 25	15 23	17 25	19 24	21 26	23 28	1 26	3 28	5 27
14	7 30	9 32	11 26	13 29	15 27	17 29	19 27	21 30	23 32	1 30	3 32	5 31
15	7 34	9 36	11 30	13 32	15 31	17 33	19 32	21 34	23 36	1 34	3 36	5 35
16	7 38	9 40	11 34	13 37	15 35	17 37	19 35	21 38	23 40	1 38	3 40	5 39
17	7 42	9 44	11 38	13 41	15 39	17 41	19 39	21 42	23 44	1 42	3 44	5 43
18	7 46	9 48	11 42	13 44	15 43	17 45	19 43	21 45	23 48	1 46	3 48	5 46
19	7 50	9 52	11 46	13 48	15 47	17 49	19 47	21 49	23 52	1 50	3 52	5 50
20	7 54	9 56	11 50	13 52	15 51	17 53	19 51	21 53	23 56	1 54	3 56	5 54
21	7 58	10 00	11 54	13 56	15 55	17 57	19 55	21 57	0 00	1 58	4 00	5 58
22	8 01	10 04	11 58	14 00	15 59	18 01	19 59	22 01	0 03	2 02	4 04	6 02
23	8 05	10 08	12 02	14 04	16 02	18 05	20 03	22 05	0 07	2 06	4 08	6 06
24	8 09	10 12	12 06	14 08	16 06	18 09	20 07	22 09	0 11	2 10	4 12	6 10
25	8 13	10 16	12 10	14 12	16 10	18 13	20 11	22 13	0 15	2 14	4 16	6 14
26	8 17	10 19	12 14	14 16	16 14	18 17	20 15	22 17	0 19	2 17	4 20	6 18
27	8 21	10 23	12 18	14 20	16 18	18 20	20 19	22 21	0 23	2 21	4 24	6 22
28	8 25	10 27	12 22	14 24	16 22	18 24	20 23	22 25	0 27	2 25	4 28	6 26
29	8 29	10 31	12 26	14 28	16 26	18 28	20 27	22 29	0 31	2 29	4 32	6 30
30	8 33		12 30	14 32	16 30	18 32	20 31	22 33	0 35	2 33	4 36	6 34
31	8 37		12 34		16 34		20 35	22 37		2 37		6 38

Although this table is correct only for the year 1960 it may be converted into a mean table for any year by subtracting 2 minutes for January and February, 3 minutes for March through July, and 2 minutes for the remainder of the year.

Compiled from the <u>American Ephemeris and Nautical Almanac</u>.

APPENDIX III
Conversion of Arc to Time

°	h m	°	h m	°	h m	°	h m	°	h m	°	h m
0	0 00	60	4 00	120	8 00	180	12 00	240	16 00	300	20 00
1	0 04	61	4 04	121	8 04	181	12 04	241	16 04	301	20 04
2	0 08	62	4 08	122	8 08	182	12 08	242	16 08	302	20 08
3	0 12	63	4 12	123	8 12	183	12 12	243	16 12	303	20 12
4	0 16	64	4 16	124	8 16	184	12 16	244	16 16	304	20 16
5	0 20	65	4 20	125	8 20	185	12 20	245	16 20	305	20 20
6	0 24	66	4 24	126	8 24	186	12 24	246	16 24	306	20 24
7	0 28	67	4 28	127	8 28	187	12 28	247	16 28	307	20 28
8	0 32	68	4 32	128	8 32	188	12 32	248	16 32	308	20 32
9	0 36	69	4 36	129	8 36	189	12 36	249	16 36	309	20 36
10	0 40	70	4 40	130	8 40	190	12 40	250	16 40	310	20 40
11	0 44	71	4 44	131	8 44	191	12 44	251	16 44	311	20 44
12	0 48	72	4 48	132	8 48	192	12 48	252	16 48	312	20 48
13	0 52	73	4 52	133	8 52	193	12 52	253	16 52	313	20 52
14	0 56	74	4 56	134	8 56	194	12 56	254	16 56	314	20 56
15	1 00	75	5 00	135	9 00	195	13 00	255	17 00	315	21 00
16	1 04	76	5 04	136	9 04	196	13 04	256	17 04	316	21 04
17	1 08	77	5 08	137	9 08	197	13 08	257	17 08	317	21 08
18	1 12	78	5 12	138	9 12	198	13 12	258	17 12	318	21 12
19	1 16	79	5 16	139	9 16	199	13 16	259	17 16	319	21 16
20	1 20	80	5 20	140	9 20	200	13 20	260	17 20	320	21 20
21	1 24	81	5 24	141	9 24	201	13 24	261	17 24	321	21 24
22	1 28	82	5 28	142	9 28	202	13 28	262	17 28	322	21 28
23	1 32	83	5 32	143	9 32	203	13 32	263	17 32	323	21 32
24	1 36	84	5 36	144	9 36	204	13 36	264	17 36	324	21 36
25	1 40	85	5 40	145	9 40	205	13 40	265	17 40	325	21 40
26	1 44	86	5 44	146	9 44	206	13 44	266	17 44	326	21 44
27	1 48	87	5 48	147	9 48	207	13 48	267	17 48	327	21 48

′	m s	″	s.
0	0 00	0	0.00
1	0 04	1	0.07
2	0 08	2	0.13
3	0 12	3	0.20
4	0 16	4	0.27
5	0 20	5	0.33
6	0 24	6	0.40
7	0 28	7	0.47
8	0 32	8	0.53
9	0 36	9	0.60
10	0 40	10	0.67
11	0 44	11	0.73
12	0 48	12	0.80
13	0 52	13	0.87
14	0 56	14	0.93
15	1 00	15	1.00
16	1 04	16	1.07
17	1 08	17	1.13
18	1 12	18	1.20
19	1 16	19	1.27
20	1 20	20	1.33
21	1 24	21	1.40
22	1 28	22	1.47
23	1 32	23	1.53
24	1 36	24	1.60
25	1 40	25	1.67
26	1 44	26	1.73
27	1 48	27	1.80

	29	30	31	32	33	34	35	36	37	38	39	40	41	42	43	44	45	46	47	48	49	50	51	52	53	54	55	56	57	58	59	60
	1.93	2.00	2.07	2.13	2.20	2.27	2.33	2.40	2.47	2.53	2.60	2.67	2.73	2.80	2.87	2.93	3.00	3.07	3.13	3.20	3.27	3.33	3.40	3.47	3.53	3.60	3.67	3.73	3.80	3.87	3.93	4.00

29	30	31	32	33	34	35	36	37	38	39	40	41	42	43	44	45	46	47	48	49	50	51	52	53	54	55	56	57	58	59	60
1 56	2 00	2 04	2 08	2 12	2 16	2 20	2 24	2 28	2 32	2 36	2 40	2 44	2 48	2 52	2 56	3 00	3 04	3 08	3 12	3 16	3 20	3 24	3 28	3 32	3 36	3 40	3 44	3 48	3 52	3 56	4 00

329	330	331	332	333	334	335	336	337	338	339	340	341	342	343	344	345	346	347	348	349	350	351	352	353	354	355	356	357	358	359	360
21 56	22 00	22 04	22 08	22 12	22 16	22 20	22 24	22 28	22 32	22 36	22 40	22 44	22 48	22 52	22 56	23 00	23 04	23 08	23 12	23 16	23 20	23 24	23 28	23 32	23 36	23 40	23 44	23 48	23 52	23 56	24 00

269	270	271	272	273	274	275	276	277	278	279	280	281	282	283	284	285	286	287	288	289	290	291	292	293	294	295	296	297	298	299	300
17 56	18 00	18 04	18 08	18 12	18 16	18 20	18 24	18 28	18 32	18 36	18 40	18 44	18 48	18 52	18 56	19 00	19 04	19 08	19 12	19 16	19 20	19 24	19 28	19 32	19 36	19 40	19 44	19 48	19 52	19 56	20 00

209	210	211	212	213	214	215	216	217	218	219	220	221	222	223	224	225	226	227	228	229	230	231	232	233	234	235	236	237	238	239	240
13 56	14 00	14 04	14 08	14 12	14 16	14 20	14 24	14 28	14 32	14 36	14 40	14 44	14 48	14 52	14 56	15 00	15 04	15 08	15 12	15 16	15 20	15 24	15 28	15 32	15 36	15 40	15 44	15 48	15 52	15 56	16 00

149	150	151	152	153	154	155	156	157	158	159	160	161	162	163	164	165	166	167	168	169	170	171	172	173	174	175	176	177	178	179	180
9 56	10 00	10 04	10 08	10 12	10 16	10 20	10 24	10 28	10 32	10 36	10 40	10 44	10 48	10 52	10 56	11 00	11 04	11 08	11 12	11 16	11 20	11 24	11 28	11 32	11 36	11 40	11 44	11 48	11 52	11 56	12 00

89	90	91	92	93	94	95	96	97	98	99	100	101	102	103	104	105	106	107	108	109	110	111	112	113	114	115	116	117	118	119	120
5 56	6 00	6 04	6 08	6 12	6 16	6 20	6 24	6 28	6 32	6 36	6 40	6 44	6 48	6 52	6 56	7 00	7 04	7 08	7 12	7 16	7 20	7 24	7 28	7 32	7 36	7 40	7 44	7 48	7 52	7 56	8 00

29	30	31	32	33	34	35	36	37	38	39	40	41	42	43	44	45	46	47	48	49	50	51	52	53	54	55	56	57	58	59	60
1 56	2 00	2 04	2 08	2 12	2 16	2 20	2 24	2 28	2 32	2 36	2 40	2 44	2 48	2 52	2 56	3 00	3 04	3 08	3 12	3 16	3 20	3 24	3 28	3 32	3 36	3 40	3 44	3 48	3 52	3 56	4 00

APPENDIX IV

Conversion of Universal Time (G.C.T.) to United States Time Zones

Universal Time	Eastern Daylight Time	Eastern Standard Time and Central Daylight Time	Central Standard Time and Mountain Daylight Time	Mountain Standard Time and Pacific Daylight Time	Pacific Standard Time
h					
0	*8 P.M.	*7 P.M.	*6 P.M.	*5 P.M.	*4 P.M.
1	*9	*8	*7	*6	*5
2	*10	*9	*8	*7	*6
3	*11 P.M.	*10	*9	*8	*7
4	0 Midnight	*11 P.M.	*10	*9	*8
5	1 A.M.	0 Midnight	*11 P.M.	*10	*9
6	2	1 A.M.	0 Midnight	*11 P.M.	*10
7	3	2	1 A.M.	0 Midnight	*11 P.M.
8	4	3	2	1 A.M.	0 Midnight
9	5	4	3	2	1 A.M.
10	6	5	4	3	2
11	7	6	5	4	3
12	8	7	6	5	4
13	9	8	7	6	5
14	10	9	8	7	6
15	11 A.M.	10	9	8	7
16	12 Noon	11 A.M.	10	9	8
17	1 P.M.	12 Noon	11 A.M.	10	9
18	2	1 P.M.	12 Noon	11 A.M.	10
19	3	2	1 P.M.	12 Noon	11 A.M.
20	4	3	2	1 P.M.	12 Noon
21	5	4	3	2	1 P.M.
22	6	5	4	3	2
23	7 P.M.	6 P.M.	5 P.M.	4 P.M.	3 P.M.

From American Ephemeris and Nautical Almanac, 1960.
*The time used is Universal Time, which differs from ordinary time by an exact number of hours as shown in the table; an asterisk denotes that the time is on the preceding day.

APPENDIX V

Suppliers of Telescope Materials and Equipment

Cave Optical Co.
4137 E. Anaheim Street
Long Beach, CA 90804

Celestron International
P.O. Box 3578
Columbia Street
Torrance, CA 90503

Chicago Optical & Supply
P.O. Box 11309
Chicago, IL 60611

P. A. Clausing
8038 Monticello Ave.
Skokie, IL 60076

Coulter Optical Co.
P.O. Box K
Idyllwild, CA 92349

Criterion Scientific Instruments
620 Oakwood Ave.
West Hartford, CT 06110

Cross Optics
569 Todd Loop
Los Alamos, NM 87544

Crown Optical, Inc.
P.O. Box 8672
Newport Beach, CA 92660

Dey Optics
Box 92313
Rochester, NY 14692

Dodd Co.
7795 W. Ridgewood Dr.
Parma, OH 44129

E & W Optical, Inc.
2420 East Hennepin Ave.
Minneapolis, MN 55413

Edmund Scientific Co.
101 East Gloucester Pike
Barrington, NJ 08007

Enterprise Optics
P.O. Box 413
Placentia, CA 92670

Evaporated Metal Films
706 Spencer Road
Ithaca, NY 14850

A. Jaegers
691 S. Merrick Road
Lynbrook, NY 11563

Kenton Engineering Corp.
641 Academy Drive
Northbrook, IL 60062

Lumicon
2111 Research Dr. #5
Livermore, CA 94550

Meade Instruments Corp.
1675 Toronto Way
Costa Mesa, CA 92626

Migiel, Thomas
4730 Highland
Downers Grove, IL 60515

North Star Telescope Co.
3542 Elm Street
Toledo, OH 43608

Optica h/c
4100 MacArthur Blvd.
Oakland, CA 94610

W. R. Parks Co.
P.O. Box 6683
Torrance, CA 90504

R.V.R. Optical Co.
P.O. Box 62
Eastchester, NY 10709

(continued on page 332)

Suppliers (For addresses, see pages 329 and 332)	Abrasives	Books	Diagonal holders	Diagonal mirrors	Drives	Drive gears	Drive motors	Dobson mirrors	Dobson telescopes	Eyepieces	Eyepiece holders
Cave Optical Co.						X				X	
Celestron International										X	
Chicago Optical Co.											
Clausing, P.A.											
Coulter Optical Co.			X						X	X	
Criterion Sci. Instr.											
Cross Optics										X	
Crown Optical			X							X	X
Dey Optics											
Dodd Co.		X								X	
E & W Optical, Inc.			X								
Edmund Scientific Co.	X	X	X		X	X				X	
Enterprise Optics											
Evaporated Metal Films											
Jaegers, A.				X							
Kenton Engr. Co.				X							
Lumicon											
Meade Instruments		X								X	
Migiel, Thomas											
North Star Telescope											
Optica h/c	X	X	X							X	
Parks, W.R.				X						X	
R.V.R. Optical										X	
Remer, N.	X									X	
S & S Optika		X								X	
Shaffer, Richard											
Sky Pub. Co.		X									
Star Instruments											
Tele-Optics		X									
Telescopics										X	
Tele-Vue Optics										X	
Tomlin Enterprises											
Tuthill, Roger W.											
Unitron Instruments, Inc.										X	X
University Optics		X								X	
VERNONscope & Co.										X	
Willmann-Bell		X									

Suppliers whose names are followed with an asterisk are those whose products have been purchased by the author or which he has had an opportunity to examine. This in no way

Filters	Finders	Hypersensitization kits	Lens kits	Lenses, achromatics	Lenses, simple	Mirror blanks	Mirror cells	Mirror coatings	Mirror kits	Mirrors, custom	Mirrors, large	Mountings, complete	Mountings, parts	Optical test kits	Setting circles	Star diagonals	Teflon pads	Telescope accessories	Telescopes, complete	Telescope tubes	Tools, laps
X								X	X		X	X						X			
X	X										X	X						X	X		
X																					X
						X															
					X				X	X											
																		X			
			X		X																
X	X								X							X		X			
X					X							X									
X																					
X																		X			
X			X	X	X		X			X	X	X						X	X		
									X												
						X												X			
			X			X		X			X										
X	X	X																			
X											X							X	X		
					X																
																		X			
X			X	X	X			X									X	X			
X	X											X						X	X		
					X						X						X	X			
X					X			X				X								X	
X																	X				
																X					
										X	X										
																		X			
							X	X	X												
	X																	X			
																		X			
	X		X														X	X			
																	X	X			
X	X																X	X			
																	X				

implies that products from other suppliers are anything but excellent, and these names and products are included in the list because of their high reputation.

N. Remer, Optics	Box 306
	Southampton, PA 18966
S & S Optika Ltd.	3855 So. Broadway
	Englewood, CO 80110
Richard D. Shaffer	228 East Pentagon
	Altadena, CA 91001
Sky Publishing Co.	49 Bay State Road
	Cambridge, MA 02238
Star Instruments	3641 E. Fox Lair Dr.
	Flagstaff, AZ 86001
Tele-Optics	2026 8th Ave. S.E.
	Calgary, Alberta, Canada
Telescopics	P.O. Box 98
	La Canada, CA 91011
Tele-Vue Optics	15 Green Hill Lane
	Spring Valley, NY 10977
Tomlin Industries	699 Easy Street
	Simi Valley, CA 93065
Tuthill, Roger W.	11 Tanglewood Lane, Box 1086
	Mountainside, NJ 07092
Unitron Instrument Co.	175 Express St.
	Plainview, NY 11803
University Optics	P.O. Box 1205
	2122 E. Delhi Road
	Ann Arbor, MI 48106
VERNONscope & Co.	Candor, NY 13743
Willmann-Bell	P.O. Box 3125
	Richmond, VA 23235

APPENDIX VI

Instructions for Silvering an 8-Inch Mirror

Materials and equipment needed:

Dishwashing detergent, either liquid or solid
2 ounces (60 grams) concentrated nitric acid, specific gravity 1.4
1 gallon distilled water
3 ounces (90 grams) ammonium hydroxide (ammonia water), specific gravity 0.90
1/2 ounce (15 grams) potassium hydroxide (caustic potash), C.P. pellet form
1 ounce (30 grams) silver nitrate, C.P.
1/2 ounce (15 grams) dextrose, U.S.P.
4 jars or beakers for silvering solutions (one must hold at least a quart; the others may be half this size or less)
thermometer
1 pair rubber gloves
stirring rod, glass, about 8 inches long
silvering tray, glass or baked enamel, approximately the size of the mirror and 2 inches deep. A photographic tray is good for this purpose, but glass is best.
small package of absorbent cotton
chamois skin, large enough to wrap around a ball of cotton to make a pad 2 inches wide, covered with a thin layer of fine dry rouge
large medicine dropper
kitchen measuring glass (ounces)
rubber suction cub, large size, with handle
pair of goggles

Directions:

1. *Scrub* all utensils (including the rubber gloves, suction cup, and anything else that may be in contact with the silvering solutions) with a strong detergent solution in hot water. Use a pad of absorbent cotton for this and apply pressure. Rinse in tap water, then scrub again.

2. Scrub the front, back, and sides of the mirror in the same way, giving particular attention to the front surface.

3. Make a pad of absorbent cotton and wrap and tie it around the end of the glass rod; use this to scrub the surface of the mirror with nitric acid. Use rubber gloves and use the silvering dish to hold the mirror during this operation. Concentrated nitric acid is a dangerous chemical: it will attack anything except glass and rubber, so keep it away from contact with skin, clothing, or anything else you value. When you have finished the job, flush everything down the sink with large amounts of water—gallons instead of ounces.

4. Rinse everything in distilled water. Use small quantities of the distilled water—six rinses with one ounce is much better than one rinse with six ounces.

5. Attach the rubber suction cup to the back of the mirror, and place the mirror, face down, in the silvering dish. Fill the dish to the level of the back of the mirror with distilled water. Silvering is usually done with the mirror face up, but it is just as efficient to do it with the mirror face down. You have better control of the mirror, the job is faster and cleaner, and the deposit of silver is just as good.

6. Make up the silvering solutions as follows:

a. Dissolve 3/4 ounce silver nitrate in 8 ounces water in the one-quart receptacle. This is the main solution.

b. Dissolve 1/8 ounce silver nitrate in 2 ounces water in a smaller dish. This is the reserve solution. Keep the remaining 1/8 ounce silver nitrate for emergencies or for your next mirror.

c. Dissolve 1/2 ounce potassium hydroxide in 8 ounces water.

d. Dissolve 1/2 ounce dextrose in 4 ounces water.

Stir each solution until each ingredient is completely dissolved. The amounts of water specified are approximate; an ounce either way won't make any difference.

7. Keep the ammonium hydroxide in a stoppered bottle; use it only as needed.

8. Allow all solutions to stand for an hour or so until they come to room temperature—close to 68 F. While you are waiting, spread out all utensils and materials on a large table so that they can be easily reached.

9. Now begin to add ammonium hydroxide *slowly* to the main silvering solution. The whole solution will turn brown. Stir constantly and continue to add ammonium hydroxide until the solution shows signs of clearing. Use the medicine dropper for further additions, stirring after each addition of a few drops. As the solution clears, add only one drop at a time. When the solution appears grayish or slightly cloudy, stop. *An excess of ammonium hydroxide will ruin all subsequent operations.* To guard against this possibility, add a few drops of the reserve silver solution until the appearance of the main solution is more cloudy than it was after you finished with the ammonium hydroxide. No harm is done if the solution contains more silver nitrate than theory permits, but too much ammonium hydroxide is fatal.

10. Add all of the potassium hydroxide solution to the main solution, *a little at a time, with constant stirring. Put on goggles for this operation. Too rapid an addition of the potassium hydroxide may cause an explosion* since a small amount of silver fulminate is formed at this stage. However, there is little likelihood of such an event if the potassium hydroxide is added slowly enough.

11. Once more add ammonium hydroxide, until the brown or black appearance of the main solution produced in the previous step begins to clear up. When the clearing starts, proceed very cautiously, again adding the ammonium hydroxide drop by drop. You are aiming for a solution which will look like weak tea, and which more than likely will contain a myriad of small black specks. Here again an excess of ammonium hydroxide is fatal to the process, so after the solution has the weak-tea appearance described above, add some more of the reserve silver nitrate solution, to be safe. The silvering solution is now complete and ready to use.

12. Lift the mirror clear of the bottom of the silvering utensil by means of the handle attached to the rubber suction cup, and pour the solution of dextrose into the distilled water surrounding the mirror.

13. Still keeping the mirror from touching any part of the silvering vessel, but making sure it is immersed in the solution, add the silvering solution to the contents of the silvering vessel. The mirror must remain suspended during the subsequent steps.

14. The next part of the process seems more like magic than chemistry. The solution will first turn a very dark brown, then a lighter brown, and after a minute or two the inside of the silvering vessel, as seen from the outside, becomes bright with a coating of silver. The silver has also begun to form on the surface of the mirror. To ensure an even coating, move the mirror very gently back and forth in the solution. The solution will begin to clear; dark specks and light brown flakes make their appearance. This clearing occurs within 5 to 7 minutes after the solutions have been mixed, and is the final critical point of the operation. If the mirror is left in the solution too long, a coating of blue-white colloidal silver called "bloom" will deposit, cutting down on the reflectivity of the surface. Watch the inside edges of the silvering vessel for this—the minute you see any signs of a white deposit, remove the mirror. In no case leave the mirror in the solution for more than 10 minutes, and 8 minutes is probably a safer margin.

15. Remove the mirror from the solution, rinse its surface with distilled water, then stand it on edge to dry. Place a piece of blotting paper against the bottom edge to soak up the final drops of water.

16. While the mirror is drying, flush all solutions down the drain, using plenty of water. Put the silvering vessel, whose inside surface will have a beautiful coating of silver, aside. You will use it in a few minutes for experimental purposes.

17. Make up a pad of chamois, using absorbent cotton as a filler, Dust its surface with the finest rouge you have. You will use this pad to burnish the mirror. Practice on the inside of the silvering vessel, using short circular strokes, until

you have learned exactly how strong its action is. Scrape the surface of the pad occasionally to remove clumps of rouge which, if allowed to accumulate, will scratch the silver surface. When you have found the best way to use the pad, transfer your efforts to the mirror, being sure the mirror is thoroughly dry before you start.

18. After burnishing is completed, the mirror will have a uniform brilliant surface. The thickness of the coating of silver may be judged by the tests suggested for aluminum coats described in Chapter 9, page 129.

Instructions for packing a mirror:

If you decide to have your mirror aluminized instead of silvering it yourself, here are some suggestions for packing it safely for shipment.

Make a rugged box, $9 \times 9 \times 2$ inches, out of thin plywood. Line it with paper to make it as dustproof as possible. Place the mirror in the bottom of the box, polished surface up. Place wedge cleats on four sides of the mirror to hold it immovably in place. The cleats should be shaped to bear only on the edge of the mirror. Use small screws to hold the cleats, and also to attach the cover so that the aluminizer may remove and replace it easily. Surround the box with excelsior or other packing material in the strongest cardboard box you can find (another wooden box is even better). All these precautions are to protect the mirror on its return trip from the aluminizers when it will be covered with a delicate coat of aluminum.

It has happened, on rare occasions, that aluminizing firms returned the wrong mirror. In order to save yourself trouble if this should happen, scratch your initials on the back of the mirror.

APPENDIX VII

The Cassegrainian Mirror and the Gaviola Test

The equivalent focal length of a Cassegrainian system can be found by multiplying the focal length of the primary mirror by what is called the amplifying factor. This amplifying factor is found by dividing the length of the cone of light formed by the secondary mirror by the length of that part of the cone of light cut off by the secondary. In Fig. 3, page 275, the amplifying factor is the quotient of a' by a; a' is also the radius of curvature of the convex secondary mirror. In practice, the secondary is made by grinding and polishing to an approximately spherical surface and then transforming the surface to a hyperboloid.

The process of testing the hyperboloid has always been a difficult and tiresome one and has involved the production of other spherical surfaces—a spherical mirror or a large optical flat. About twenty years ago the astronomer Enrique Gaviola, then working at the La Plata Observatory in Argentina, evolved a method for testing the hyperboloidal secondary of the 32-inch La Plata mirror.

In essence, this method consists of measuring zones on the secondary and comparing them with theoretical measurements. The equipment required consists of:

1. A monochromatic light source (slit) and knife edge.
2. A reading glass or other convergent lens slightly larger than the mirror to be tested. The lens must have a focal length smaller than that of the mirror to be tested. The errors in the lens are unimportant since, by the way the test is conducted, they cancel themselves out.
3. A small flat mirror such as the diagonal from a Newtonian reflector.
4. A zonal screen placed in front of the mirror.

The departure of a hyperboloidal surface from that of a sphere is called its eccentricity (e). The eccentricity of the secondary can be found from the formula:

$$e = \frac{A + 1}{A - 1}$$

where A is the amplifying factor (see above). Once the eccentricity is known, the theoretical zonal measurements are given by the formula:

$$Z_t = \frac{e^2 r^2}{2R}$$

where Z_t = longitudinal aberration of each zone or, more simply, the variation in knife-edge position
    e = eccentricity
    r = radius of the zone measured
    R = radius of curvature of the secondary mirror

The physical set-up for making the actual measurements is not difficult, although the test must be done in two parts. First, the knife-edge and slit must be placed as nearly coincident to each other as possible. The lens, screen, and mirror are set up in that order with a surface of the lens touching the screen on one side and the convex surface of the mirror on the other. The distance of the three units from the slit is approximately equal to the focal length of the lens. The light from the slit now passes through the lens and screen openings, then is reflected from the mirror surface back through the screen and lens to the knife-edge.

The zero point from which readings are measured is the point where the central zone in the screen darkens evenly. The other zones are then tested, one by one, in terms of the knife-edge distance from the knife-edge. These readings are tabulated. They do not represent the Z values found above, of course, since the lens errors affect their validity.

The second part of the test eliminates the lens errors. In this part, slit, knife-edge, lens, and screen are left exactly in the zero position, but the convex mirror is removed. It is replaced by the small flat mirror, which is moved away from the screen to a position (whose distance from the screen will again be approximately equal to the focal length of the lens) where the zero point is duplicated at the knife-edge. The knife-edge remains stationary throughout this part of the test. The flat mirror is now moved to positions which duplicate the readings found in the previous test. The shift in position of the mirror from its zero position for each zone now represents the actual measurement for the longitudinal aberration of that zone. These measurements are compared to the theoretical values for Z_t found above, and appropriate polishing techniques are applied until the values coincide closely.

This outline of the Gaviola Test is necessarily brief. The original paper[1] should be consulted for a more complete description.

[1]See footnote on page 314.

APPENDIX VIII

Table of Sagitta

Mirror Diam. (inches)	Focal Ratio	Radius of Curvature	Sagitta[1] $S = r^2/2R$	Sagitta[2] $S = R - \sqrt{R^2 - r^2}$	Difference
6	f/8	96	0.0469	0.0469	0.0000
6	f/6	72	0.0625	0.0625	0.0000
6	f/4	48	0.0937	0.0938	0.0001
6	f/2	24	0.1875	0.1882	0.0007
8	f/7	112	0.0714	0.0714	0.0000
8	f/6	96	0.0833	0.0833	0.0000
8	f/4	64	0.1250	0.1251	0.0001
8	f/2	32	0.2500	0.2510	0.0010
10	f/4	80	0.1563	0.1564	0.0001
10	f/2	40	0.3125	0.3137	0.0012
12	f/4	96	0.1875	0.1877	0.0002
12	f/2	48	0.3750	0.3765	0.0015

The second formula given above is the more accurate in determining the depth of curve, or sagitta, of a mirror. The table shows that it makes little difference, for mirrors of moderate focal length, which formula is used. But for short focal length mirrors, the difference becomes increasingly important. If you plan to grind an RFT of f/4, a difference in sagitta of 0.0001 inch for a 6-inch mirror is translated into a difference of radius of curvature of nearly 1 1/2 inches.

Adapted from a table by Allan Mackintosh. Used by permission.

APPENDIX IX

Table of Tangents

Degree	Tangent	Degree	Tangent	Degree	Tangent
0	.000	31	.601	61	1.80
1	.017	32	.625	62	1.88
2	.035	33	.649	63	1.96
3	.052	34	.675	64	2.05
4	.070	35	.700	65	2.14
5	.087	36	.727	66	2.25
6	.105	37	.754	67	2.36
7	.123	38	.781	68	2.48
8	.141	39	.810	69	2.61
9	.158	40	.839	70	2.75
10	.176	41	.869	71	2.90
11	.194	42	.900	72	3.08
12	.213	43	.933	73	3.27
13	.231	44	.966	74	3.49
14	.249	45	1.000	75	3.73
15	.268	46	1.03	76	4.01
16	.287	47	1.07	77	4.33
17	.306	48	1.11	78	4.70
18	.325	49	1.15	79	5.14
19	.344	50	1.19	80	5.67
20	.364	51	1.23	81	6.31
21	.384	52	1.28	82	7.12
22	.404	53	1.33	83	8.14
23	.424	54	1.38	84	9.51
24	.445	55	1.43	85	11.4
25	.466	56	1.48	86	14.3
26	.488	57	1.54	87	19.1
27	.510	58	1.60	88	28.6
28	.532	59	1.66	89	57.3
29	.554	60	1.73	90	∞
30	.577				

APPENDIX X

Standard Pipe Fittings

Nominal inside diameter of pipe (inches)	$\frac{1}{2}$	$\frac{3}{4}$	1	$1\frac{1}{4}$	$1\frac{1}{2}$	2	$2\frac{1}{2}$	3	$3\frac{1}{2}$	4	$4\frac{1}{2}$	5
Actual inside diameter of pipe (inches)	.622	.824	1.049	1.380	1.610	2.067	2.469	3.068	3.548	4.026	4.506	5.047
Outside diameter of pipe (inches)	.840	1.050	1.315	1.660	1.900	2.375	2.875	3.500	4.000	4.500	5.000	5.563
Threads per inch	14	14	$11\frac{1}{2}$	$11\frac{1}{2}$	$11\frac{1}{2}$	$11\frac{1}{2}$	8	8	8	8	8	8
Weight per foot (pounds)	.850	1.130	1.678	2.272	2.717	3.652	5.793	7.575	9.108	10.790	12.538	14.617
Tap drill size (inches)	$\frac{11}{16}$	$\frac{29}{32}$	$1\frac{1}{8}$	$1\frac{15}{32}$	$1\frac{23}{32}$	$2\frac{3}{16}$	$2\frac{9}{16}$	$3\frac{3}{16}$	$3\frac{11}{16}$	$4\frac{3}{16}$	—	—

APPENDIX XI

Another Method of Making Setting Circles

The method of making setting circles of a specified size given on page 223 of the text can, of course, be adapted to other sizes. But difficulties of measurement arise in the adaptation, and so we give an alternate method here. This excellent idea is quoted direct from an article by Allan Mackintosh in his *Maksutov Club Notes,* and is used with his specific permission.

"As soon as one has completed and used a first telescope, the question arises how to save time in finding objects which are invisible to the naked eye. Sweeping the sky with a finder can be a very frustrating occupation.

"The answer is, of course, setting circles and the immediate following question is how to make them without a dividing head or dividing engine, both of which are likely to cost much more than the total budget the family treasurer is likely to allow to the hapless T.N. for several years.

"My solution to the problem is a combination of the ideas of many people gleaned through the years, with a few techniques of my own thrown in. Having seen some dozens of setting circles, of both amateur and professional manufacture, I have come to the conclusion that the chief drawbacks to most of them are lack of visibility on a dark night and cumulative errors in the circles themselves. Many amateurs' circles are too small, and even when the divisions are large enough to be seen easily in dim light, they become quite invisible on the long-hoped-for moonless night of good seeing. Everyone knows the disastrous effect a flashlight has on eyes, even when covered with red paper; it takes at least ten to fifteen minutes to get maximum sensitivity back.

"Circles should be as large as the telescope will take without appearing absurd. Turn or shape up a couple of circles of the required diameter out of 1/2" plywood and then cut a couple of strips of Lucite, plexiglass, or some other transparent material—they should be about 1 1/2" wide and a little longer than the circumference of the plywood discs. The Lucite should be 1/16" thick for large circles, less for small ones. Wrap them around the discs and cut and file the ends to a good butt join.

"Stretch the piece of Lucite which is to become the declination circle along the side of a table and secure it to the table with masking tape. Get yourself a yardstick, or better still a 36" steel scale if it is available, and a T-square. Set the T-square at one end of the plastic and arrange the yardstick at an angle to the Lucite. If the Lucite is longer than the yardstick, fasten the yardstick along the edge of the table and the Lucite at an angle to it—in this case nick the marks on the edge of the Lucite and scribe them later with a square.

"Using the T-square tight against the side of the Lucite, scribe lines using every inch on the yardstick. I use a 1/4" lathe tool ground to a 60° angle for my scriber and use it with the flat side leading; if you secure the Lucite to the table lengthwise with the piece of masking tape, this will give you a very good guide for uniform length of the lines.

"Having done this, you will find that you have a strip of Lucite scribed into 36 divisions; errors will be in inverse ratio to the care with which you have done the work, but in any case they will not be cumulative. Each of the 36 divisions will represent 10 on the Dec. circle and most people want their circles divided into degrees. Get a 6" scale divided into 1/10ths inch (they are easily obtainable) and make a subsidiary scale in 1/32" aluminum by the same angular method; carefully divide each of the 36 divisions into 10 subdivisions; errors here depend on your eyesight, but again they are not cumulative.

"Cover the side on which you have made the scribe-marks with 2" masking tape and cut out numbers of the required size from newsprint or whatever comes to hand (not your wife's cookbook!). You don't need to cut around the number, just cut it from the paper in a small square, thus ☐2. Stick the squares to the masking tape in the proper positions *face down;* if you have used newsprint, you will be able to see an outline of the number through the paper and, of course, the number will be backwards, though as the circle reads from 1 to 90 and back to 1, the order from left to right or right to left does not matter. With a sharp knife (Exacto,

Sears' workshop knife, or something else which is *really* sharp) and a steel straight-edge, cut through the paper and the masking tape around the edges of the numbers right down to the Lucite. When all the numbers have been cut, gently strip off the masking tape from the end; since the numbers have been cut clear from the rest of the tape, they will remain on the Lucite.

"Get a spray can of dead-black paint and spray the Lucite black on the side the numbers are stuck on. When the paint is dry, lift off the numbers carefully, using the point of a knife to get under the masking tape. Properly done, this will result in a black scale of the proper length for your plywood disc with the numbers in clear Lucite. Carefully scrape out the paint from the scribe-marks and you will find that looking through the unpainted side, you have a very excellent job with the numbers the right way round. Put it aside.

"The hour-angle scale is treated in the same way except that every 1 1/2" is used on the yardstick, giving 24 divisions instead of 36. Subdivisions can be made by the same angular method using 1/8" on your scale—this will give you 12 sub-divisions per 1 1/2", or 5 minutes of time per hour, a very convenient number. Also on this scale the numbers should read 1 to 24 from right to left, so that when you read it from the painted side, they will read from left to right.

"You will find that the edges of your T-square are of plastic, but it is a sim-ple matter to fix it up with a steel strip so that the scriber will not bite into the plastic. If you have to borrow one, it is best to use a round scriber to make a light line, using the lathe-tool scriber against a short steel straightedge to deepen the line. Carefully done, there will be no lessening in accuracy and you may de-cide to use this method in any case rather than mutilate your expensive straight-edge.

"Next glue your scales carefully around your plywood discs, painted side in-wards, and finish off the joint by putting on a piece of Scotch tape covered with two or three coats of transparent plastic varnish.

"I will not go into the details of boring the plywood discs for the telescope shafts, or the details of the 'innards' of the H-A disc, but ask you to remember that there should be some method of locking the discs to the shafts and that the H-A disc should have some arrangement so that you can slip and lock it at will.

"Next, get a couple of ruby pea or flashlight bulbs; the ruby ones are the best, but a clear bulb covered with dark red paint works almost as well. These are mounted in small housings which can be built of 1/32" aluminum, or a suitable can, or what have you. This has a slot cut in the top large enough so that one (or two) numbers are always visible in the opening. The pointer is made of a piece of wire about 1/16" thick stretched across the opening; the wire blacks out the lines and is easily seen from this, though you may want to paint it with luminous paint. The inside of the housing is painted dead-black.

"The ruby light, of course, gives a dim red light through the transparent fig-ures and scribe marks on the circle. This is easily sufficient to be read very eas-ily and the darker the surroundings, the better the figures stand out, but the light given off is not sufficient to affect the sensitivity of the retina at all. A little ex-perimentation will give you the right intensity for your eyes and if you want to be very fancy, a small rheostat inserted in the wiring of the bulbs will allow you to vary the intensity as you wish.

"One further note about the dimensions of the scribed lines. Make them as fat as you have patience for—1/32" is good. Do not make them too long as length of the lines leads to confusion in reading them. I like the following for my circles:

Dec. circle	Hour angle circle
10° lines 1/2 inch	hour lines 1/2 inch
5° lines 1/4 inch	1/2 hour lines 1/4 inch
degree lines 1/8 inch	5 minute lines 1/8 inch

APPENDIX XII

Frequencies for Radio Time Signals

Station Call Letters	Frequencies of Transmission (in megacycles)
WWV (National Bureau of Standards)	2.5, 5.0, 10.0, 15.0, 20.0, 25.0
WWVH	5.0, 10.0, 15.0, 20.0
GIC37	17.685
GUAM	17.530
LOL	17.180
NSS	5.870, 9.4250, 12.8040, 17.050
NBA	17.127
NTG	17.055
GBZ	19.6
TQG5	13.873
CBU	7.335

APPENDIX XIII

Constellations Visible in the Northern Hemisphere

Name	Common Name	Approximate Date When on the Meridian
Andromeda	Andromeda	November 10
Aquarius	The Water Carrier	October 10
Aquila	The Eagle	August 30
Aries	The Ram	December 10
Auriga	The Charioteer	January 30
Boötes	The Herdsman	June 15
Cancer	The Crab	March 15
Canes Venatici	The Hunting Dogs	May 20
Canis Major	The Great Dog	February 15
Canis Minor	The Little Dog	March 1
Capricornus	The Goat	September 20
Cassiopeia	Cassiopeia	November 20
Cepheus	The King	October 15
Cetus	The Whale	November 30
Coma Berenices	Berenice's Hair	May 15
Corona Borealis	The Northern Crown	June 30
Corvus	The Crow	May 10
Crater	The Cup	April 25
Cygnus	The Swan	September 10

Name	Common Name	Approximate Date When on the Meridian
Delphinus	The Dolphin	September 15
Draco	The Dragon	July 20
Equuleus	The Horse	September 20
Eridanus	The River	January 5
Gemini	The Twins	February 20
Hercules	Hercules	July 25
Hydra	The Sea Serpent	April 20
Lacerta	The Lizard	October 10
Leo	The Lion	April 10
Leo Minor	The Little Lion	April 10
Lepus	The Hare	January 25
Libra	The Scales	June 20
Lyra	The Lyre	August 15
Monoceros	The Unicorn	February 20
Ophiuchus	The Serpent Bearer	July 25
Orion	Orion	January 25
Pegasus	The Flying Horse	October 20
Perseus	Perseus	December 25
Pisces	The Fish	November 10
Sagitta	The Arrow	August 30
Sagittarius	The Archer	August 20
Scorpius	The Scorpion	July 20
Scutum	The Shield	August 15
Serpens Caput	The Serpent's Head	June 30
Serpens Cauda	The Serpent's Tail	August 5
Sextans	The Sextant	April 5
Taurus	The Bull	January 15
Triangulum	The Triangle	December 5
Ursa Major	The Big Dipper, or The Great Bear	April 20
Ursa Minor	The Little Dipper, or The Little Bear	June 25
Virgo	The Virgin	May 25
Vulpecula	The Fox	September 10

The Rhyme of the Zodiac

The Ram, the Bull, the heavenly Twins,
And next the Crab, the Lion shines,
The Virgin, and the Scales.
The Scorpion, Archer, and the Goat
The Man who pours the Water out,
And Fish with glittering tails.

APPENDIX XIV

Some Bright Stars of the Northern Heavens (from 40° North Latitude)

Star	Constellation in Which it Appears	Magnitude	RA		Decl. (1960)	
			h	m	°	′
Adhara	Canis Major	1.6	06	57	−28	55
Aldebaran	Taurus	1.1	04	33	+16	26
Algenib	Perseus	1.9	03	21	+49	43
Alhena	Gemini	1.9	06	35	+16	26
Alioth	Ursa Major	1.7	12	52	+56	11
Alkaid	Ursa Major	1.9	13	46	+49	31
Alnila.n	Orion	1.8	05	34	−01	14
Alnitak	Orion	1.8	05	38	−01	58
Alphard	Hydra	2.2	09	36	−08	29
Altair	Aquila	0.9	19	49	+08	46
Antares	Scorpius	1.2	16	27	−26	21
Arcturus	Boötes	0.2	14	14	+19	23
Bellatrix	Orion	1.7	05	23	+06	19
Betelgeuse	Orion (Variable)	0-1	05	53	+07	24
Capella	Auriga	0.2	05	14	+45	58
Castor	Gemini	1.6	07	32	+31	59
Deneb	Cygnus	1.3	20	40	+45	08
Denebola	Leo	2.2	11	47	+14	48
Dubhe	Ursa Major	1.9	11	01	+61	58
El Nath	Taurus	1.8	05	24	+28	34
Fomalhaut	Pisces Austrinus	1.3	22	55	−29	50
Kaus Australis	Sagittarius	1.9	18	21	−34	24
Murzim	Canis Major	2.0	06	21	−17	56
Procyon	Canis Minor	0.5	07	37	+05	20
Regulus	Leo	1.3	10	06	+12	10
Rigel	Orion	0.3	05	13	−08	15
Sargas	Scorpius	2.0	17	34	−42	58
Shaula	Scorpius	1.7	17	31	−37	05
Sirius	Canis Major	−1.6	06	43	−16	40
Spica	Virgo	1.2	13	23	−10	57
Vega	Lyra	0.1	18	35	+38	45
Wezen	Canis Major	2.0	07	06	−26	20

Adapted from the American Ephemeris and Nautical Almanac, 1960.

APPENDIX XV

Astronomical Symbols

☿ Mercury

♀ Venus

⊕ Earth

♂ Mars   ☌ Conjunction. Where two bodies have the same right ascension but not necessarily the same declination.

♃ Jupiter

♄ Saturn

♅ Uranus   ☍ Opposition. Where two bodies differ 180° in right ascension.

♆ Neptune

♇ Pluto   ☐ Quadrature. Where two bodies differ 90° in right ascension.

☉ Sun

● New Moon

☽ First Quarter

○ Full Moon

☾ Last Quarter

APPENDIX XVI

Greek Alphabet

A	α	Alpha	H	η	Eta	N	ν	Nu	T	τ	Tau
B	β	Beta	Θ	θ	Theta	Ξ	ξ	Xi	Y	ν	Upsilon
Γ	γ	Gamma	I	ι	Iota	O	o	Omicron	Φ	ϕ	Phi
Δ	δ	Delta	K	κ	Kappa	Π	π	Pi	X	χ	Chi
E	ϵ	Epsilon	Λ	λ	Lambda	P	ρ	Rho	Ψ	ψ	Psi
Z	ζ	Zeta	M	μ	Mu	Σ	σ	Sigma	Ω	ω	Omega

APPENDIX XVII

Suggested Films and Exposure Times for Astrophotography

Apparatus	Object	Exposure	Films
Fixed camera; EFL up to 12″	Star trails Meteors Aurora Constellations Bright comets	10–30 min. 10 ″ 1–5 ″ 1–5 ″ 1–2 ″	Plus-X Verichrome Pan Ilford Pan F Ilford FP4 Variable Speed AgfaPan Kodak 2415 Kodacolor 25 or 64 Ektachrome 64
Guided camera EFL up to 20″	Dim comets Deep sky objects Milky Way	15–30 min.	103a series, depending on color of objects Kodacolor 400 Fujicolor 400 Ektachrome 200 or 400 Fujichrome 400
Unguided telescope at prime focus	Moon Planets Sun	$\frac{1}{500}$–1 sec.	
Guided telescope at prime focus; EFL up to 100″	Moon, planets Deep sky objects (clusters, nebulae, etc.)	10–30 min	Fast color films listed above Tri-X alone or in cold camera Hypered Kodak 2415 or colored films
Guided telescope with eyepiece projection; EFL over 100″	Close-ups of moon, planets	1–2 sec.	

EFL = Effective focal length

Suggested developers: D6, D19, Microphen, Acufine, MWP-2, others used according to directions listed on their packaging with respect to contrast, definition, etc.

APPENDIX XVIII

Astronomical Organizations

Although the information found in atlases, books, magazines, and other publications devoted to astronomy will suggest many uses to which you can put your telescope, it is quite probable that you will learn more from contact with other telescope owners than from any other source. There is probably an amateur astronomical group in your own neighborhood, or within driving distance, and you will have a fine opportunity to exchange ideas and information by joining it. *Sky and Telescope* magazine periodically lists such organizations as well as the names and addresses of the people to contact for admission to membership. Usually the only qualification for joining one of these groups is your own interest in astronomy. The smaller clubs are set up to provide close contact with people whose interest in some aspect of observing is similar to your own, and the exchange of ideas is often stimulating and rewarding in improving your techniques. Larger groups, such as the Amateur Astronomers Association in New York City, provide this kind of contact as well as lectures, workshop facilities, field trips, informative leaflets about current astronomical events, and many other services to its members.

On a much larger scale, but providing the same kind of stimulation, are the national organizations. We list several below.

The Astronomical Society of the Pacific

Membership in this large and well-organized group is open to anyone interested in astronomy. The central organization has its offices at 675 Eighteenth Avenue, San Francisco, California.

Member clubs include those from the western part of the United States, most of which are located along the Pacific Coast. The purpose of the Society is to promote the science of astronomy by spreading astronomical information through the media of publications, lectures, meetings, and excursions. It publishes an eight-page monthly leaflet as well as bimonthly *Publications,* each of which present excellent articles on various phases of astronomy and current astronomical research. For the most part, these articles are nontechnical in nature. Members are entitled to attend all lectures, meetings, and excursions, and also receive the publications.

The Western Amateur Astronomers

Made up of many amateur clubs in the western United States, this expanding organization is affiliated with the Astronomical Society of the Pacific. It holds a yearly three-day convention for the exchange of ideas in all phases of astronomy, whether it be mirror-making, amateur observational techniques, or the relationship of the amateur and professional astronomer. Such exchanges are accomplished by means of lectures, star parties, field trips to large observatories, and symposia conducted by amateur and professional groups working together.

Although the principal impetus for amateur telescope making and amateur astronomy in general originated in the east, largely through the efforts of the original group at Springfield, Vermont, there is much activity in the western areas of the United States, and many mirror-making techniques and ideas have originated there.

The American Association of Variable Star Observers (AAVSO)

This group goes back to 1911, when observations of variable stars were needed to supplement the work of the observatories. Such observations are still very useful, and membership in the Association is open to anyone over sixteen years of age who has a small or medium-sized telescope. Headquarters are at 187 Concord Avenue, Cambridge, Massachusetts 02138. Many famous astronomers have been connected with the AAVSO, among them Leon Campbell, for many years Recorder of the Association. His equally famous son, Leon Campbell, Jr., is the coordinator of the "Moonwatch" program. The noted astronomer and writer, Margaret Mayall, now serves as Recorder.

The American Meteor Society

The purpose of this group is to encourage and promote the observation of meteor trails, as well as to increase interest in astronomy in general. Its membership includes professional and amateur astronomers alike. The Society publishes a journal called *Meteoritics,* a record of its activities and reports. The observation and photography of meteors is a fascinating hobby, well worth the attention of the amateur. Communications concerning membership may be sent to the State University of New York, Geneseo, New York 14454.

The Association of Lunar and Planetary Observers

This association represents another opportunity for the amateur to use his telescope for an exciting and useful purpose. The moon and planets present a source of almost limitless investigation, especially in these days when satellites are supplementing the work of telescopes as lunar probes. Further information may be obtained from Walter Haas, Box 3 AZ, University Park, Las Cruces, New Mexico, 88003.

There are, of course, many avenues the amateur can follow in his pursuit of astronomy. Although there are relatively few opportunities for positions as professional astronomers, room can always be made for those who have the ability and interest to succeed in the profession. Percival Lowell, the founder of the great observatory which bears his name, acquired his original interest as an amateur. So, also, did Russell W. Porter, the man who supplied the enthusiasm and skill in the first organization of amateurs at Springfield, Vermont, and who is now famous for his beautifully executed drawings of the great Palomar telescope.

APPENDIX XIX

Suggested Additional Reading

Books on Telescope Making

Brown, Sam	*All About Telescopes*	Sky Publishing Co., Cambridge, Mass.
Ingalls, A. G., ed.	*Amateur Telescope Making* *Amateur Telescope Making, Advanced* *Amateur Telescope Making, Book III*	Scientific American Publishing Co., New York, N.Y.
Moore, Patrick	*Astronomical Telescopes and Observatories for Amateurs*	W. W. Norton, New York, N.Y.
Muirden, James	*Amateur Astronomer's Handbook*	Cassell, London, Eng.
Paul, Henry	*Telescopes for Star Gazing*	Amphoto, New York, N.Y.
Sidgwick, J. B.	*The Amateur Astronomer's Handbook*	Thomas Y. Crowell, New York, N.Y.
Thompson, Allyn J.	*Making Your Own Telescope*	Sky Publishing Co., Cambridge, Mass.

Observational Astronomy

Abell, George	*Realm of the Universe*	Holt, Rinehart & Winston, New York, N.Y.
Alter, Dinsmore	*Pictorial Guide to the Moon*	Thomas Y. Crowell, New York, N.Y.
Beatty, O'Leary, Chalkin, eds.	*The New Solar System*	Sky Publishing Co., Cambridge, Mass.
Gehrels, Tom	*Jupiter*	University of Arizona Press, Tucson, Ariz.
Glasby, John	*Variable Star Observer's Handbook*	W. W. Norton, New York, N.Y.
Gamow, George	*A Star Called the Sun*	Viking Press, New York, N.Y.
Hawkins, G. S.	*Meteors, Comets, and Meteorites*	McGraw-Hill, New York, N.Y.
Moore, Patrick	*Observational Facts and Feats* *A Guide to Mars*	Guiness Superlatives, London, Eng. Macmillan, New York, N.Y.
NASA	*Rings of Saturn SP343*	Government Printing Office, Washington, D.C.
Paul, Henry	*Outer Space Photography*	Amphoto, New York, N.Y.

Sherrod, P. Clay	*Complete Manual of Amateur Astronomy*	Prentice-Hall, Englewood Cliffs, N.J.

Charts and Reference Works

American Ephemeris		Government Printing Office, Washington, D.C.
Howard, Neale E.	*Telescope Handbook and Star Atlas*	Thomas Y. Crowell, New York, N.Y.
Mallas, John, and Evered Kreimer	*Messier Album*	Sky Publishing Co., Cambridge, Mass.
Norton, Arthur P.	*Star Atlas*	Sky Publishing Co., Cambridge, Mass.
Rey, H. A.	*The Stars*	Houghton Mifflin, Boston, Mass.
Royal Astronomical Society	*The Observer's Handbook*	Toronto, Can.
Tirion, Wil	*Sky Atlas 2000*	Sky Publishing Co., Cambridge, Mass.

Magazines

Astronomy	Monthly	Astromedia Corp., 625 E. St. Paul Ave., P.O. Box 92788, Milwaukee, Wisc. 53202
Sky & Telescope	Monthly	Sky Publishing Co., 49 Bay State Road, Cambridge, Mass. 02138
Telescope Making	Quarterly	Astromedia Corp., 625 E. St Paul Ave., Milwaukee, Wisc. 53202

NOTE: All of the books listed above may be obtained directly from their publishers or through the following:

Willmann-Bell, Inc. P.O. Box 3125, Richmond Virginia 23235.
Sky Publishing Co., Address listed above.
Herbert A. Luft, 69-11 229th Street, Box 91, Oakland Gardens, N.Y. 11364.

Index